Tutoring Integral Calculus

concepts needed to help others learn

S. Gill Williamson

Preface

From 1965 to 1991 I was a professor in the Department of Mathematics at the University of California, San Diego (UCSD). I taught many calculus classes large and small during this period. In 1991 I transferred to the Department of Computer Science and Engineering - my calculus teaching days were over.

Recently (2011), cleaning out some files, I came across a long lost typewritten handout that I used to give to students who officially or unofficially wanted to tutor for my integral calculus classes. I had fun rereading this "tutors' guide" so I decided to redo it in LaTeX and bring it up to date with respect to online resources now regularly used by students.

This material assumes that as a prospective integral calculus tutor you have mastered the standard undergraduate level differential and integral calculus courses. The most common conceptual and pedagogical pitfalls of tutoring integral calculus are discussed along with worked exercises.

S. Gill Williamson, 2012
http : \www.cse.ucsd.edu\ ~ gill

Table of Contents

Chapter 1

Integrals as Antiderivatives

1.1 Integration as the inverse of differentiation

This material assumes that you have had a course in calculus through integral calculus and want to tutor (or home tutor) students who are studying integral calculus for the first time. We focus on developing the skill and intuition to prepare yourself to face a small group of students who are stuck or, worse yet, not stuck and full of questions. We start with a review of basics. In our discussion, we have in mind real valued functions defined for all real numbers or intervals of real numbers just like the ones you studied in calculus up to this point.

The symbol "$\frac{d}{dx}$" is used to mean "the derivative of." Thus, $\frac{d}{dx} \sin(x)$ means "the derivative of $\sin(x)$." Correspondingly, we use the symbol " $\int$ " to mean "the integral of" or "antiderivative of." Thus, we write $\int 2x \cos(x^2) = \sin(x^2)$ to mean the integral of $2x \cos(x^2)$ is $\sin(x^2)$: $\frac{d}{dx} \sin(x^2) - 2x \cos(x)$.

(1.1.1)   **Figure : Integral vs. Derivative**

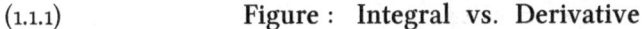

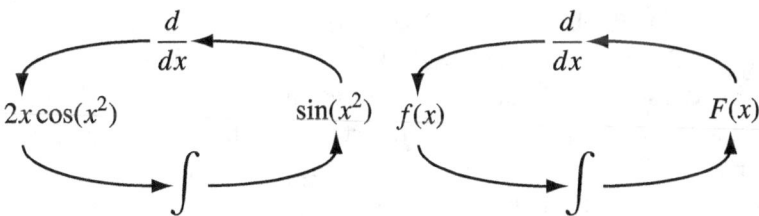

If $f(t)$ is specified as a function of t then $\int f(t)$ is commonly understood to be a function $F(t)$ such that $\frac{d}{dt} F(t) = f(t)$. But what if we walk into an abandoned classroom and see " $\int 2 = ?$" on the blackboard. What was the variable?

If it was x, then the answer is $2x$. If the variable was t, then the answer is $2t$. To avoid this and related confusions, the notation for integrals or antiderivatives is written $\int f(x)\,dx$ or $\int f(t)\,dt$. Thus, if we had seen "$\int 2\,dt =$?" then the answer would have been $2t$. If we had seen "$\int 2\,dx =$?" then the answer would have been $2x$.

$$\boxed{F(x) \text{ and } F(x) + C \textbf{ \textit{have the same derivative.}}}$$

There is another simple but important observation about integrals. As we have noted, $\int 2x\cos(x^2) = \sin(x^2)$, which means that $\frac{d}{dx}\sin(x^2) = 2x\cos(x^2)$. But, of course, $\frac{d}{dx}(\sin(x^2) + 10) = 2x\cos(x^2)$ also. In fact, $\frac{d}{dx}(\sin(x^2) + C) = 2x\cos(x^2)$ for any constant function C. This fact is sometimes incorporated into the notation for integrals by writing

$$(1.1.2) \qquad\qquad \int 2x\cos(x^2)\,dx = \sin(x^2) + C.$$

This notation is intended to remind us that there are infinitely many functions with derivative $2x\cos(x^2)$ and they all differ by a constant function. Once this observation has been made and we understand what we are talking about, it is quite all right to write simply

$$\int 2x\cos(x^2)dx = \sin(x^2).$$

We understand in this latter notation that $\sin(x^2)$ is a representative from an infinite class of antiderivatives for $2x\cos(x^2)$, and all of the rest are obtained by adding a constant function to $\sin(x^2)$.

$$\boxed{H'(x) = 0 \implies H(x) = C, \textbf{\textit{ a constant function.}}}$$

The fact that all antiderivatives of a given function $f(x)$ differ by a constant is a subtle idea. Suppose we have two functions, $F(x)$ and $G(x)$, such that $F'(x) = G'(x) = f(x)$. Let $H(x) = F(x) - G(x)$. Then $H'(x) = F'(x) - G'(x) = f(x) - f(x) = 0$. To claim that $F(x)$ and $G(x)$ *differ by a constant function* is the same as claiming that $H(x) = F(x) - G(x)$ *is* a constant function. This means that the statement that "any two antiderivatives $F(x)$ and $G(x)$ of $f(x)$ differ by a constant" is the same as the statement that "any function $H(x)$ with derivative function the zero function must be a constant function." This latter statement has strong intuitive appeal.

Suppose $H'(x) = 0$ for all x. Let's try to draw the graph of such an $H(x)$. Suppose $H(0) = 2$, for example. Put your pencil at the point $(0, 2)$ and try to imagine what the graph is like near this point. If, in going right or left, you draw the graph with the slightest bit of slope up or down you will construct points on the graph where $H'(x)$ is not 0. You're stuck at $H(x) = 2$ and must draw the graph of the constant function 2. For more advanced courses in mathematical analysis it is essential that this intuitive idea be given a precise analytical formulation.

> ### Constant Functions Can Wear Many Disguises

Here is another complication. Suppose John decides that

$$\int 2\sin(2x)\,dx = -\cos(2x)$$

and suppose that Mary decides that

$$\int 2\sin(2x)dx = 2\sin^2(x).$$

If they are both right (and they are in this case) then $2\sin^2(x)$ and $-\cos(2x)$ must differ by a constant (i.e., $2\sin^2(x) - (-\cos(2x)) = 2\sin^2(x) + \cos(2x)$ is a constant function). If you know your basic trigonometric identities, then you will recognize that this is true and, in fact, $2\sin^2(x) + \cos(2x) = 1$. Thus, just because two integrals $F(x)$ and $G(x)$ for $f(x)$ *must* differ by a constant doesn't mean that they are *easily recognizable* as differing by a constant.

1.2 Properties of the integral

The most basic property of integrals is "linearity." This property, which we have already used, is stated in Theorem 1.2.1.

Theorem 1.2.1. *Let $f(x)$ and $g(x)$ be functions and α and β numbers. Then*

$$\int (\alpha f(x) + \beta g(x))dx = \alpha \int f(x)dx + \beta \int g(x)dx.$$

Proof. This follows directly from the definition of the integral together with linearity of $\frac{d}{dx}$. Let $F(x)$ and $G(x)$ be the antiderivatives of $f(x)$ and $g(x)$. Then $\frac{d}{dx}(\alpha F(x) + \beta G(x)) = \alpha f(x) + \beta g(x)$, which means, by definition of the integral, that

$$\int (\alpha f(x) + \beta g(x))dx = \alpha F(x) + \beta G(x).$$

Substituting $F(x) = \int f(x)\,dx$ and $G(x) = \int g(x)\,dx$ gives the result. $\qquad\square$

Your students' main task in computing antiderivatives or integrals will be to develop systematic ways to reduce new problems to ones they have already solved. Theorem 1.2.1 is a start in this direction. We have already noticed that $\int 2x \cos(x^2)\, dx = \sin(x^2)$ and $\int \sin(2x) = \sin^2(x)$. Thus we can evaluate an integral such as $\int (2\pi x \cos(x^2) + 25 \sin(2x))\, dx$ by using Theorem 1.2.1:

$$\pi \int 2x \cos(x^2)\, dx + 25 \int \sin(2x)\, dx = \pi \sin(x^2) + 25 \sin^2(x)\,.$$

> ### Your students already know many integrals.

As they begin to compute more integrals, your students will use the rule of Theorem 1.2.1 automatically. They will discover that they already know many integrals! Every differentiation formula they have memorized gives rise to a corresponding integration formula:

$$\frac{d}{dx} \sin(x) = \cos(x) \quad \text{becomes} \quad \sin(x) = \int \cos(x) dx$$

$$\frac{d}{dx} \ln(|x|) = 1/x \quad \text{becomes} \quad \ln|x| = \int \frac{1}{x}\, dx$$

$$\frac{d}{dx} x^n = nx^{n-1} \quad \text{becomes} \quad x^n = \int nx^{n-1} dx\,.$$

This latter integral has been memorized by millions of calculus students as

$$\int x^n dx = \frac{x^{n+1}}{n+1} \quad \text{for} \quad n \neq -1\,.$$

For $n = -1$, $\int x^{-1} dx = \ln|x|$.

> Show: $\frac{d}{dx} \sec(x) = \sec(x)\tan(x) \qquad \frac{d}{dx} \csc(x) = -\csc(x)\cot(x)$

By applying Theorem 1.2.1, you can compute integrals such as

$$\int (34 \sin(x) + 23x^3 + 45 \sec^2(x)) dx = -34 \cos(x) + 23\frac{x^4}{4} + 45 \tan(x)\,.$$

(1.2.2)   **Chain rule in reverse**

To really get started on the problem of computing integrals, your students must learn how to do the *chain rule* in reverse. In particular,

(1.2.3) $\dfrac{d}{dx} f(g(x)) = f'(g(x))g'(x) \implies f(g(x)) = \displaystyle\int f'(g(x))g'(x) dx\,.$

Let's start with an easy example. Consider $\int \cos(x^2)(2x)\,dx$. It appears that $f'(x) = \cos(x)$ and $g(x) = x^2$ here. This is determined by guessing or by inspection with the aid of past experience. Thus, $\int \cos(x^2)(2x)\,dx = \int f'(g(x))g'(x)\,dx = f(g(x)) = \sin(x^2)$. We had to guess that $g(x) = x^2$ was the proper choice since the function g is not explicitly mentioned. The way we make such a guess is by knowing our derivative formulas well. We know that $\frac{d}{dx}x^2 = 2x$. As our eyes scan the expression $\cos(x^2)(2x)$, we spot the pair x^2 and its derivative $2x$. This is the clue that prompts us to try $g(x) = x^2$.

In addition to guessing correctly that $g(x) = x^2$ in the previous example, we had to know how to integrate $f'(x) = \cos(x)$ to get $f(x) = \sin(x)$. As another example, let's try to calculate $\int \ln(x^2)(2x)\,dx$.

Again, we see the pair $g(x) = x^2$ and $g'(x) = 2x$. Thus $f'(x) = \ln(|x|)$ so that $f'(g(x))g'(x) = \ln(x^2)(2x)$ (note that $|x^2| = x^2$). To compute $f(g(x))$, we must compute the integral $f(x)$ of $f'(x) = \ln(|x|)$. $f(x) = x\ln(|x|) - x$ is the answer which you can check by differentiating. From this fact, we conclude that $\int \ln(x^2)(2x)\,dx = (x^2)\ln(x^2) - x^2$. Check this statement by computing the derivative of the expression on the right.

Another complication that occurs is seen in the following two integrals:

$$\int x\cos(x^2)\,dx =? \quad \text{and} \quad \int 5x\ln(x^2)\,dx =? \,.$$

In these integrals, we see the $g(x) = x^2$ all right, but the $g'(x) = 2x$ is not there. Instead, we see x in the first integral and $5x$ in the second integral. We know that for any number α and any function $h(x)$, $\int \alpha h(x)\,dx = \alpha \int h(x)\,dx$. Thus,

$$\int x\cos(x^2)\,dx = \int (1/2)(2x)\cos(x^2)\,dx = (1/2)\int \cos(x^2)(2x)\,dx\,.$$

We have already discovered that $\int \cos(x^2)(2x)\,dx = \sin(x^2)$, so we have $\int x\cos(x^2)\,dx = (1/2)\sin(x^2)$.

$$\boxed{\text{Show: } \frac{d}{dx}\tan(x) = \sec^2(x) \quad \frac{d}{dx}\cot(x) = -\csc^2(x)}$$

(1.2.4) **Differential notation**

Certain notations of calculus are designed to help in the task of applying the chain rule in reverse. We know that $\int \cos(x)dx = \sin(x)$. There is, of course, nothing special about the x here:

$$\int \cos(t)\,dt = \sin(t)\,, \quad \int \cos(\tau)d\tau = \sin(\tau)\,, \quad \int \cos(A)dA = \sin(A)\,,$$

and

$$\int \cos(JUNK)d(JUNK) = \sin(JUNK).$$

The general rule is that if $F(x)$ is an integral or an antiderivative of $f(x)$) (i.e., $F'(x) = f(x)$) then

$$\int f(JUNK)d(JUNK) = F(JUNK)$$

where JUNK stands for anything for which these formulas make sense. JUNK can be quite complicated. For example

$$\int \cos\left(\frac{\ln(x)e^x \tan(\sin(x))}{x^5 + 2x^3 + 5x + 1}\right) d\left(\frac{\ln(x)e^x \tan(\sin(x))}{x^5 + 2x^3 + 5x + 1}\right) = \sin\left(\frac{\ln(x)e^x \tan(\sin(x))}{x^5 + 2x^3 + 5x + 1}\right).$$

In this formula

$$JUNK = \left(\frac{\ln(x)e^x \tan(\sin(x))}{x^5 + 2x^3 + 5x + 1}\right).$$

Here is a more formal presentation:

Definition 1.2.5 (differential notation). Let $g(x)$ be a function of x with $\frac{dg}{dx} = g'(x)$. Then define $dg = g'(x)\,dx$. This is the *differential notation for the derivative of g*. The term dg is called the *differential of g* and the term dx is called the *differential of x*.

We can state our above discussion as a theorem:

Theorem 1.2.6. *Let $F(x)$ be such that $F'(x) = f(x)$ and let $g(x)$ be a function with differential $dg = g'(x)\,dx$. Then,*

$$\int f(g)\,dg = F(g).$$

Proof. For any function $h(x)$ we have, by the definition of the integral, $h(x) = \int h'(x)\,dx = \int dh$. Applying this to the chain rule (where $F'(x) = f(x)$) gives

$$F(g(x)) = \int d(F(g(x))) = \int F'(g(x))\,g'(x)\,dx = \int f(g(x))\,dg.$$

Thus $\int f(g)\,dg = F(g)$. $\qquad\qquad\square$

How We Make Up Confusing Problems

You should show your students how easy it is to make up confusing problems that "illustrate" the method of substitution. For example, let $\cos(x)$ be $f(x)$.

Thus, $F(x) = \sin(x)$. Next we pick any function $g(x)$ that we can differentiate. Let's take $g(x) = \ln(x^2 + 1)$. Then, $g'(x) = \frac{2x}{x^2+1}$ or $dg = \frac{2x}{x^2+1}\,dx$. Substituting into the identity $\int f(g)\,dg = F(g)$ we obtain

$$\int \cos(\ln(x^2 + 1))\,d(\ln(x^2 + 1)) = \sin(\ln(x^2 + 1)).$$

But this is the same as

$$\int \cos(\ln(x^2 + 1))\frac{2x}{x^2 + 1}\,dx = \sin(\ln(x^2 + 1))$$

which is (almost) the same as

$$\int \frac{\cos(\ln(x^2 + 1))}{x + 1/x}\,dx = ?\,.$$

The following Exercises 1.3 are designed to give you practice with the method "integration by substitution" discussed in the previous paragraphs. In these exercises, you are to find functions $f(x)$ and $g(x)$ such that the given problem can be reduced to computing $\int f(g)\,dg$.

1.3   Exercises

Integration by substitution

Evaluate the following (for the appropriate values of x). Solutions follow.

Exercise 1.3.1. $\displaystyle\int x\sqrt{x^2 + 2}\,dx = \quad$ ($f(x) = x^{1/2}$ and $g(x) = x^2 + 2$)

Exercise 1.3.2. $\displaystyle\int \frac{x^{1/4}}{5 + x^{5/4}}\,dx = \quad$ ($g(x) = 5 + x^{5/4}$)

Exercise 1.3.3. $\displaystyle\int \frac{\sqrt{x} - 1}{\sqrt{x}}\,dx =$

Exercise 1.3.4. $\displaystyle\int \frac{8x}{1 + e^2 x}\,dx = \quad$ ($g(x) = 1 + e^2 x$)

Exercise 1.3.5. $\displaystyle\int \frac{x\cos\sqrt{5x^2 + 1}}{\sqrt{5x^2 + 1}}\,dx = \quad$ ($f(x) = \cos(x), g(x) = (5x^2 + 1)^{1/2}$)

Exercise 1.3.6. $\displaystyle\int \sin^{3/4} 2x\,\cos 2x\,dx = \quad$ ($g(x) = \sin(2x)$)

Exercise 1.3.7. $\displaystyle\int \tan^2 3t\,\sec^2 3t\,dt = \quad$ ($f(t) = t^2$)

Exercise 1.3.8. $\displaystyle\int x\sec(x^2)\tan(x^2)\,dx =$

13

Exercise 1.3.9. $\displaystyle\int \frac{2\ln^3(|x|)}{x}\, dx =$

Exercise 1.3.10. $\displaystyle\int \frac{6x}{(x^2+9)^3}\, dx =$ $(g(x) = x^2 + 9)$

Exercise 1.3.11. $\displaystyle\int \frac{e^{\log_2(x)}}{x}\, dx =$ $(g(x) = \log_2(x)$ or try $\log_2(x) = \log_2(e)\log_e(x))$

Exercise 1.3.12. $\displaystyle\int \frac{1}{\sin\theta}\, d\theta =$ $(g(\theta) = \cos(\theta))$

Solutions to Exercises 1.3

Solution 1.3.1: $\int x\sqrt{x^2+2}\, dx = \int x g^{1/2}\, dx$ where $g = (x^2 + 2)$. Thus, $dg = 2x\, dx$ so $x\, dx = dg/2$. Evaluate $\int g^{1/2} dg/2 = (1/2)(2/3)g^{3/2}$. The answer is $(1/3)(x^2 + 2)^{3/2}$. All antiderivatives of $x\sqrt{x^2+2}$ are of the form $(1/3)(x^2 + 2)^{3/2} + C$, where C is a constant function. We generally omit the constant C as previously discussed.

Solution 1.3.2: $\displaystyle\int \frac{x^{1/4}}{5 + x^{5/4}}\, dx = \frac{4}{5}\int (5 + x^{5/4})^{-1}\, d(5 + x^{5/4}) = \frac{4}{5}\ln(5 + x^{5/4}).$

Solution 1.3.3: Beware of this type of problem! No tricky substitutions are required:
$$\int \frac{\sqrt{x}-1}{\sqrt{x}}\, dx = \int (1 - x^{-1/2})\, dx = 1 - 2x^{1/2}.$$

Solution 1.3.4: Let $g = (1 + e^2 x)$. Then $x = e^{-2}(g-1)$, $dx = e^{-2}\, dg$, and
$$\int \frac{8x}{1 + e^2 x}\, dx = 8e^{-4}\int \frac{g-1}{g}\, dg = 8e^{-4}\left(\int dg + \int g^{-1}dg\right) = 8e^{-4}(g + \ln(|g|)).$$

Now set $g = (1 + e^2 x)$. Note the similarity to Solution 1.3.3.

Solution 1.3.5: $\displaystyle\int \frac{x\cos\sqrt{5x^2+1}}{\sqrt{5x^2+1}}\, dx = \int \cos\left((5x^2+1)^{1/2}\right)\, d\left(\frac{1}{5}(5x^2+1)^{1/2}\right) =$

$\displaystyle\frac{1}{5}\int \cos\left((5x^2+1)^{1/2}\right)\, d(5x^2+1)^{1/2} = \frac{1}{5}\sin\left((5x^2+1)^{1/2}\right).$

Solution 1.3.6: $\int \sin^{3/4} 2x\, \cos 2x\, dx = \frac{1}{2}\int \sin^{3/4}(2x)\, d\sin(2x) = \frac{2}{7}\sin^{7/4}(2x).$

Solution 1.3.7: $\int \tan^2 3t \; \sec^2 3t \, dt = \int \tan^2(3t) \, (1/3) \, d\tan(3t) = (1/9) \tan^3(3t)$.

Solution 1.3.8: $\int x \sec(x^2) \tan(x^2) dx = (1/2) \int d(\sec(x^2)) = (1/2) \sec(x^2)$.

Solution 1.3.9: $\int \dfrac{2 \ln^3(|x|)}{x} \, dx =$ Let $g(x) = \ln(|x|)$.

Solution 1.3.10: $\int \dfrac{6x}{(x^2 + 9)^3} \, dx =$ Let $g(x) = x^2 + 9$.

Solution 1.3.11: $\int \dfrac{e^{\log_2(x)}}{x} \, dx =$ Let $g(x) = \log_2(x)$ or directly substitute $\log_2(x) = \log_2(e) \log_e(x)$. The computations go as follows: Using $\log_2(x) = \log_2(e) \log_e(x)$ we get

$$\int \frac{e^{\log_2(x)}}{x} \, dx = \int \frac{e^{\log_2(e) \ln(x)}}{x} \, dx = \int \frac{x^{\log_2(e)}}{x} \, dx = \int x^{\log_2(e)-1} \, dx = \frac{x^{\log_2(e)}}{\log_2(e)} \, .$$

Substituting $g(x) = \log_2(x)$ we get $\int \dfrac{e^{\log_2(x)}}{x} \, dx = \ln(2) \int e^g \, dg = \ln(2)e^g = \ln(2)e^{\log_2(x)} = \ln(2)e^{\log_2(e) \ln(x)} = \ln(2)x^{\log_2(e)}$. Note that $\log_2(e) = 1/\ln(2)$.

Solution 1.3.12: $\int \dfrac{1}{\sin \theta} d\theta =$ Let $g(\theta) = \cos(\theta)$ and use $\sin^2(\theta) = 1 - \cos^2(\theta)$:

$$\int \frac{\sin(\theta)}{\sin^2(\theta)} \, d\theta = \int \frac{\sin(\theta)}{1 - \cos^2(\theta)} \, d\theta = -\int \frac{dg}{1 - g^2} = -1/2 \left(\int \frac{dg}{1 + g} + \int \frac{dg}{1 - g} \right).$$

The answer is $(-1/2)(\ln |1 + g| - \ln |1 - g|)$ which can be written

$$\frac{1}{2} \ln |1 - g| - \frac{1}{2} \ln |1 + g| = \ln(1 - g)^{1/2} - \ln(1 + g)^{1/2} = \ln \left(\frac{(1 - g)^{1/2}}{(1 + g)^{1/2}} \right).$$

We drop the absolute values because both $1 - g$ and $1 + g$ are nonnegative as $g = \cos(\theta)$. Using the half-angle formulas $(1 - \cos(\theta))^{1/2} = 2^{1/2} |\sin(\theta/2)|$ and $(1 + \cos(\theta))^{1/2} = 2^{1/2} |\cos(\theta/2)|$, we can write

$$\int \frac{1}{\sin \theta} \, d\theta = \ln \left(2^{1/2} |\sin(\theta/2)| \right) - \ln \left(2^{1/2} |\cos(\theta/2)| \right) = \ln(|\tan(\theta/2)|).$$

1.4 Using computer resources

At the time of writing this book, a Web search on "google directory science math" brings up a list of categories that includes "Calculus." Clicking on "Calculus" brings you to another list of categories that includes "Software" where you will find a link to "The Integrator" (http://integrals.wolfram.com/). The Wolfram Integrator is easy to use and, entering our Exercise 1.3.12 (in the form $1/\sin(x)$), yields $\log(\sin(x/2)) - \log(\cos(x/2))$ as the integral. In the notation used here, "log" refers to the natural logarithm "ln". This expression, $\ln(\sin(x/2)) - \ln(\cos(x/2))$, is one of several forms for the integral that we discovered in our solution to 1.3.12. We found the equivalent form

$$\ln\left(2^{1/2}|\sin(x/2)|\right) - \ln\left(2^{1/2}|\cos(x/2)|\right) = \ln\left(|\sin(x/2)|\right) - \ln\left(|\cos(x/2)|\right).$$

Note that $1/\sin(x)$ is also written $(\sin(x))^{-1}$ or $\sin^{-1}(x)$ (consistent with $\sin^2(x)$ for $(\sin(x))^2$). You will also find that some websites use $\sin^{-1}$ for $\arcsin(x)$. We prefer the latter notation but use both as both are common.

Using the online integrator is a very easy way to solve Exercise 1.3.12. Maybe you should advise your students to stop studying calculus and just rely on mathematical software! Mathematical software can be a tremendous help to even the most sophisticated mathematician or scientist. However, there is no replacement for understanding the ideas and techniques of calculus. As a minor point, if you understand the ideas of calculus you will note that the given solution, $\ln(\sin(x/2)) - \ln(\cos(x/2))$, would be better written as

$$\ln\left(|\sin(x/2)|\right) - \ln\left(|\cos(x/2)|\right)$$

where we have included absolute values. This is a trivial observation, but it might save you from making an annoying mistake in certain contexts. Someone who uses mathematical software without knowing the theory behind what they are doing is like someone who drives a car without the slightest idea of how the car works or how to fix minor problems.

As an example, suppose we do a web search for $\int \sec(x)\,dx$. At the first site, I found $\int \sec(x)\,dx = 2\tanh^{-1}(\tan(x/2))$. On two other sites I found

$$\int \sec(x)\,dx = \ln|\sec(x) + \tan(x)| \quad \text{and} \quad \int \sec(x)\,dx = \tanh^{-1}(\sin(x)).$$

This seemed to be all of the variations until I hit upon

$$\int \sec(x)\,dx = (1/2)\ln|1 + \sin(x)| - (1/2)\ln|1 - \sin(x)|.$$

You will have students who will come up with all of these variations and more. It is worthwhile to go through these variations and understand them. The same pattern will be repeated on any problem where web resources are used.

(1.4.1) **Variations on** $\displaystyle\int \sec(x)\,dx$

As an example of how to analyze the various solutions your students might bring to you from web searches, we go through some possibilities for $\int \sec(x)\,dx$.

1. $\int \sec(x)\,dx = \ln|\sec(x) + \tan(x)|$: However they obtained this solution, ask your students to specify a domain for the function $F(x) = \ln|\sec(x) + \tan(x)|$. A natural domain is $-\pi/2 < x < \pi/2$. On this domain, $\sec(x) + \tan(x) = (1 + \sin(x))/\cos(x)$ is positive. They should verify that $F'(x) = \sec(x)$ by direct differentiation. They should graph (using a graphing program or calculator) the function $F(x)$ together with $\sec(x)$ and verify visually that $F'(x)$ is approximately $\sec(x)$. If $F(x)$ was gotten from a table or program, you should ask for (or provide) a derivation using basic principles. For example,

$$\int \sec(x)\,dx = \int \frac{\sec(x)(\sec(x) + \tan(x))\,dx}{\sec(x) + \tan(x)} = \int \frac{d(\sec(x) + \tan(x))}{\sec(x) + \tan(x)}.$$

2. $\int \sec(x)\,dx = (1/2)\ln|1 + \sin(x)| - (1/2)\ln|1 - \sin(x)|$: Assume that $-\pi/2 < x < \pi/2$. A simple direct derivation is as follows:

$$\int \sec(x)\,dx = \int \frac{\cos(x)\,dx}{1 - \sin^2(x)} = \frac{1}{2}\int \frac{d(1 + \sin(x))}{1 + \sin(x)} - \frac{1}{2}\int \frac{d(1 - \sin(x))}{1 - \sin(x)}.$$

3. $\int \sec(x)\,dx = \tanh^{-1}(\sin(x))$: Here, $\tanh^{-1}(x) \equiv \operatorname{arctanh}(x)$ is the inverse of the hyperbolic tangent. Recall that $\frac{d}{du}\tanh^{-1}(u) = \frac{1}{1-u^2}$ where $|u| < 1$. Starting as in the previous example, with $-\pi/2 < x < \pi/2$, we get:

$$\int \sec(x)\,dx = \int \frac{\cos(x)\,dx}{1 - \sin^2(x)} = \int \frac{d\sin(x)}{1 - \sin^2(x)} = \tanh^{-1}(\sin(x)).$$

4. $\int \sec(x)\,dx = 2\tanh^{-1}(\tan(x/2))$: We again use the fact that $\frac{d}{du}\tanh^{-1}(u) = \frac{1}{1-u^2}$ where $|u| < 1$. We take $-\pi/2 < x < \pi/2$ so that $-\pi/4 < x/2 < \pi/4$ and $|\tan(x/2)| < 1$. The calculations go like this:

$$\int \sec(x)\,dx = \int \frac{dx}{\cos(2(x/2))} = \int \frac{dx}{\cos^2(x/2) - \sin^2(x/2)}.$$

17

Multiply numerator and denominator in the last integral by $\sec^2(x)$:

$$\int \frac{\sec^2(x/2)\,dx}{1 - \tan^2(x/2)} = 2 \int \frac{d\,\tan(x/2)}{1 - \tan^2(x/2)} = 2\tanh^{-1}(\tan(x/2)).$$

5. $\int \csc(x)\,dx = 2\coth^{-1}(\cot(x/2))$ by analogy, doesn't it? Some student is sure to come up with this conjecture. Have the student compute the derivative of $2\coth^{-1}(\cot(x/2))$ for $0 < x < \pi/2$. The answer is again $\sec(x)$. This means that for $0 < x < \pi/2$

$$2\coth^{-1}(\cot(x/2)) = 2\tanh^{-1}(\tan(x/2)) + C.$$

In fact, the constant $C = 0$. From the definitions, if $0 < z < 1$ then

$$\tanh^{-1}(z) = \ln\left(\frac{1+z}{1-z}\right) = \ln\left(\frac{1/z+1}{1/z-1}\right) = \coth^{-1}(1/z).$$

To show $C = 0$ for $0 < x < \pi/2$, take $z = \tan(x/2)$ so $1/z = \cot(x/2)$. Web resources won't make your tutoring job any easier. Sometimes you will have to say, "I don't know."

1.5 Additional techniques of integration

(1.5.1) **Integration by Parts**

The method of substitution, uses the "chain rule in reverse." The method of "integration by parts" is the "product rule in reverse."

We have $(fg)' = f'g + fg'$ thus $f(x)g(x) = \int f'(x)g(x)\,dx + \int f(x)g'(x)\,dx$. Stated as a theorem, we have the following:

Theorem 1.5.2. *Integration by parts:* Let $h(x)$ be a function which we have written as $h(x) = f(x)g'(x)$ for some choice of f and g'. Then

(1.5.3) $$\int h(x)\,dx \equiv \int f(x)g'(x)\,dx = f(x)g(x) - \int f'(x)g(x)\,dx.$$

Proof. Write $(fg)' = f'g + fg'$ as $fg' = (fg)' - f'g$ and integrate. $\square$

This is a trivial theorem but a very useful trick of integration. The basic idea is to start with a function $h(x)$ expressed analytically in a nice fashion (e.g., $h(x) = x(2 + 3x)^{-1/2}$). We want to compute $\int h(x)\,dx$. To do so, we write $h(x) = f(x)g'(x)$ where we know how to compute $f'(x)$ and $\int g'(x)\,dx$ (both answers in a desirable analytic form). In this case, take $f(x) = x$ and $g'(x) = (2 + 3x)^{-1/2}$. Thus, $f'(x) = 1$ and $g(x) = \int g'(x)\, dx = (2/3)(2 + 3x)^{1/2}$.

Applying 1.5.3, we get

$$\int x(2+3x)^{-1/2}\, dx = x\left((2/3)(2 + 3x)^{1/2}\right) - \int g'(x)\, dx = (x-1)(2/3)(2+3x)^{1/2}.$$

Some students are helped by the following tabular form of 1.5.3:

(1.5.4)   Figure : Integration by Parts Table

$$\int f(x)\, g'(x) = f(x)\, g(x) - \int f'(x)\, g(x)$$

$f(x)$	$g(x)$
$f'(x)$	$g'(x)$

Diagonal product = top product - other diagonal

1.6   Exercises

Integration by parts

Verify the following identities using integration by parts:

Exercise 1.6.1. $\int \arctan(x) = x\arctan(x) - (1/2)\ln(1 + x^2)$

Hint:

$\arctan(x)$	x
$\dfrac{1}{1 + x^2}$	1

Exercise 1.6.2. $\int x\arctan(x) = (\frac{x^2}{2})\arctan(x) - \int(\frac{x^2}{2})(\frac{1}{1+x^2})\, dx = ?$

Hint:

$\arctan(x)$	$\dfrac{x^2}{2}$
$\dfrac{1}{1 + x^2}$	x

Exercise 1.6.3. $\int \sin(\ln(x))\, dx = x\sin(\ln(x)) - \int \cos(\ln(x))\, dx = \,?$

$\sin(\ln(x))$	x
$\dfrac{\cos(\ln(x))}{x}$	1

$\longrightarrow$

$\cos(\ln(x))$	x
$\dfrac{-\sin(\ln(x))}{x}$	1

Apply integration by parts twice and then solve for $\int \sin(\ln(x))\, dx$.

Exercise 1.6.4. $\int \sec^3\, dx = \sec(x)\tan(x) - \int \sec(x)\tan^2(x)\, dx = \,?$

Recall: $\tan^2 = \sec^2 - 1$

$\sec(x)$	$\tan(x)$
$\sec(x)\tan(x)$	$\sec^2(x)$

$$\int \sec(x)\tan^2(x)\, dx = \int \sec(x)(\sec^2(x)-1)\, dx = \int \sec^3(x)\, dx - \int \sec(x)\, dx.$$

Use any of the forms 1.4.1 for $\int \sec(x)\, dx$. Substitute the $\int \sec(x)\tan^2(x)\, dx$ result into the first expression of this exercxise and solve for $\int \sec^3(x)\, dx$.

(1.6.5)   **Integral tables and online math resources**

You should be familiar with online tables of integrals and mathematics software (including graphing software), and you should encourage your students to learn to use these resources. The online resources you use should be free to the students. There are free tables of integrals available in pdf format. It is helpful if you and your students have the same table of integrals, use the same grapher, integrator, etc. Entries in the table of integrals expressed in terms of parameters or defined recursively need to be explained to the students in terms of usage and proof.

Our first example expresses the integral in terms of real numbers α and β, $\alpha^2 \neq \beta^2$. The numbers α and β are *parameters* in the integration formula.

$$(1.6.6) \qquad \int \sin(\alpha x)\sin(\beta x)\, dx = \frac{\sin((\alpha - \beta)x)}{2(\alpha - \beta)} - \frac{\sin((\alpha + \beta)x)}{2(\alpha + \beta)}.$$

Specialized versions of 1.6.6 are found in every calculus book. Recall that

$$(1.6.7) \qquad \cos(\theta \pm \gamma) = \cos(\theta)\cos(\gamma) \mp \sin(\theta)\sin(\gamma)$$

and hence $2\sin(\alpha x)\sin(\beta x) = \cos(\alpha x - \beta x) - \cos(\alpha x + \beta x)$. Substituting this latter identity into the left hand side of identity 1.6.6 and integrating $\cos(\alpha - \beta)x$ and $\cos(\alpha + \beta)x$ gives the result.

Next we give an example of a *recursively* specified integral parametrized by integers m and n, $m \neq -n$. This identity is in all standard tables of integrals:

$$\text{(1.6.8)} \qquad \int \cos^m(x) \sin^n(x)\, dx =$$

$$\frac{\cos^{m-1}(x)\sin^{n+1}(x)}{m+n} + \frac{m-1}{m+n} \int \cos^{m-2}(x)\sin^n(x)\, dx.$$

To prove 1.6.8 we use integration by parts 1.5.4:

$\cos^{m-1}(x)$	$\dfrac{\sin^{n+1}(x)}{n+1}$
$(m-1)\cos^{m-2}(x)(-\sin(x))$	$\cos(x)\sin^n(x)$

We obtain the following recursive identity

$$\text{(1.6.9)} \qquad \int \cos^m(x) \sin^n(x)\, dx =$$

$$\frac{\cos^{m-1}(x)\sin^{n+1}(x)}{n+1} + \frac{m-1}{n+1} \int \cos^{m-2}(x)\sin^{n+2}(x)\, dx.$$

Recursion 1.6.9 holds the sum of the powers of $\cos(x)$ and $\sin(x)$ constant while reducing the power of $\cos(x)$ by two in the integral. For most applications, the identity 1.6.9 would be as good as 1.6.8. The idea is to reduce the complexity of the integral by repeated application of 1.6.9. This reduction will only work for certain values of m. For example, if $m = 3$ (or any positive odd integer), the final integral on the right will be a power of $\sin(x)$ times $\cos(x)$ which is easy to integrate.

We need some additional steps to prove 1.6.8. First, note that

$$\text{(1.6.10)} \qquad \int \cos^{m-2}(x) \sin^{n+2}(x)\, dx = \int \cos^{m-2}(x) \sin^n \sin^2(x)\, dx =$$

$$\int \cos^{m-2}(x) \sin^n(x)\, dx - \int \cos^m(x) \sin^n(x)\, dx \quad (\text{use } \sin^2(x) = 1 - \cos^2(x)).$$

Using 1.6.10, make the following substitution in 1.6.9.

$$\int \cos^{m-2}(x) \sin^{n+2}(x)\, dx = \int \cos^{m-2}(x) \sin^n(x)\, dx - \int \cos^m(x) \sin^n(x)\, dx$$

and solve the resulting equation for $\int \cos^m(x) \sin^n(x)\, dx$. The result is 1.6.8.

The integration formula 1.6.8 is but one of many that are fairly general in description. The proofs of these identities can be tricky, as we have seen. A good student will often ask a professor or tutor to explain why these integral identities are true. If you don't know it is best to say so. These identities have been used and refined over hundreds of years. After years of teaching calculus, I have derived most of them – but usually not on the spot as a result of a student's question.

To summarize, using online resources can be challenging to the tutor and students, but the effort is well worthwhile. The student's learning experience is greatly enhanced and the powerful tools of calculus become even more useful. We have included a Table of Integrals at the end of this manuscript. This table is from the book *Top-down Calculus*, by S. Gill Williamson which is available free on the author's website (see Preface).

Chapter 2

Fundamental Theorem and Definite Integrals

2.1 The fundamental theorem of calculus

In the this chapter, we present some of the theory of integration. Failure to understand this small bit of theory has resulted in some of the most serious cases of bad advice that tutors have given to my students.

For numbers p and q, we use the notation $[p, q]$ to denote the set (or *interval*) of all x between p and q (all x such that $p \leq x \leq q$, if $p \leq q$, or all x such that $p \geq x \geq q$, if $p > q$). The notation (p, q) denotes $[p, q]$ minus the two endpoints.

Given a continuous function $f(x)$ on an interval $[s, t]$, how hard is it to find an antiderivative (integral) $F(x)$ of $f(x)$? Problems like those in Exercises 1.6, where we start with the function $f(x)$ specified analytically (i.e., by some expression) and want the answer $F(x)$ also to be specified analytically, can get very difficult. However, if we start with $f(x)$ in graphical form and don't mind getting $F(x)$ in graphical form also, then finding $F(x)$ is, in principle, very easy. We now guide you through Figure 2.1.1 to help you understand this graphical approach.

(2.1.1)   **Figure : Signed Area Basics** $F(x) = S_a^x(f)$

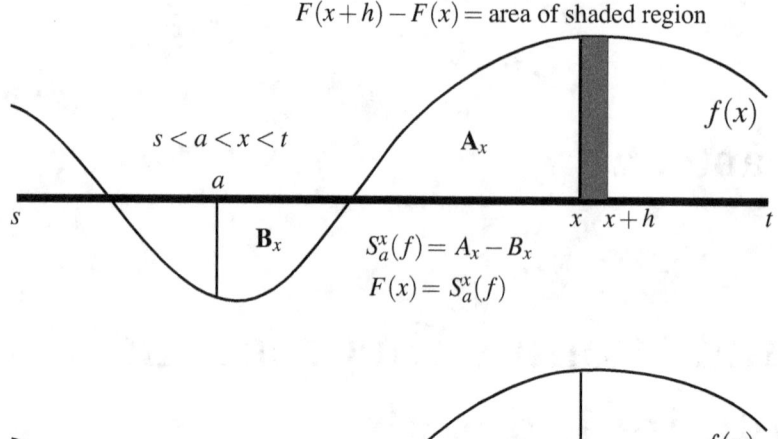

1. **The signed area function** $F(x) = S_a^x(f)$: Start with the upper half of Figure 2.1.1, and look at the graph of the function $f(x)$. Using this graph of f (where $a < x$) we define a second function, $F(x) = S_a^x(f)$, called the *signed area function of $f(x)$ with base point a*. For the x shown on the horizontal axis, consider the region $\mathbf{R}_x = \mathbf{A}_x \cup \mathbf{B}_x$ bounded by the interval $[a, x]$ (on the horizontal axis) and the graph of f. $\mathbf{R}_x$ is the union of two sub-regions: $\mathbf{A}_x$, the points on or above the horizontal axis, with area A_x, and $\mathbf{B}_x$, the points on or below the axis, with area B_x. We define $S_a^x(f) = A_x - B_x$. In the reverse case where $x < a$, shown in the lower half of Figure 2.1.1, we define $S_a^x = B_x' - A_x'$. Note that for any a and b, $S_a^b = -S_b^a$.

2. **Properties of** $F(x) = S_a^x(f)$: In Figure 2.1.1, we didn't sketch a graph of $F(x)$. See Figure 2.1.4 if your curious about what the graph of F looks like. Look again at the upper graph in Figure 2.1.1. The area of the shaded region bounded by the graph of f and the interval $[x, x + h]$ is *exactly* $F(x + h) - F(x)$ (definition of signed area) and *approximately* $h \times f(x)$. Thus, $(F(x + h) - F(x))/h \approx f(x)$ for small h, and

$$\lim_{h \to 0} \frac{F(x + h) - F(x)}{h} \approx \lim_{h \to 0} f(x) = f(x)$$

Thus, the derivative $F'(x) = f(x)$ and $F(x) = S_a^x(f)$ is an antiderivative

24

(integral) of $f(x)$. This result is called the Fundamental Theorem of Calculus. The "proof" just given is sloppy and gives only the basic intuition.

Figure 2.1.4 is an expanded version of Figure 2.1.1. In Figure 2.1.4, we take $a = -3$ and graph $F(x) = S_a^x$ (dashed curve). The numbers at the bottom

$$0, +0.5, 0, -1.17, -2.34, -2.84, -2.34, -1.04, +0.66, +2.36, +3.66, +4.16$$

correspond to the sequence $F(-5), F(-4), F(-3), F(-2), \ldots, F(6)$. The differences, $F(i+1) - F(i)$, $i = -5, -4, \ldots 6$, are also shown on the graph. They read $-0.5, +0.5, -1.17, +1.17, -0.5, +0.5, \ldots$. Your students should check that the $F'(x) = f(x)$ (approximately) for various values of x. The following table might help.

$F'(-4) = 0$	$F'(-3) = -1$	$F'(0) = 0$	$F'(1) = 1$	$F'(5) = 1$
$f(-4) =$	$f(-3) =$	$f(0) =$	$f(1) =$	$f(5) =$

Next we give a more careful definition of the signed area function, a statement of the theorem, and a less sloppy proof.

Definition 2.1.2 ($S_a^x(f)$). Let $f(x)$ be a continuous function on some interval $[s, t]$, $s < t$, and let a be a number $s < a < t$. For all x in $(s, t) = \{x : s < x < t\}$, the region of the plane bounded by the graph of f and the interval $[a, x]$ of the horizontal axis is divided into two sub-regions: $\mathbf{A}_x$, the region on or above the horizontal axis with area A_x, and $\mathbf{B}_x$, the region on or below the horizontal axis, with area B_x. The *signed area function*, $S_a^x(f)$ of f with base point a, is defined by $S_a^x(f) = A_x - B_x$ if $a < x$, and $S_a^x(f) = B_x - A_x$ if $x < a$. $S_a^x(f) = 0$ if $a = x$.

Theorem 2.1.3 (Fundamental theorem). *Let f be defined and continuous for all x in the interval $[s, t]$, $s < t$, and let a be a number in (s, t). Let $F(x) = S_a^x(f)$, x in $[s, t]$, be the signed area function of f with base point a. For every point x in (s, t), the function $F(x)$ has a derivative and $\frac{d}{dx} F(x) = f(x)$. Thus the signed area function is an antiderivative or integral of $f(x)$ on (s, t).*

Proof. (Theorem 2.1.3) In general, for a number x and for any $h > 0$, let $I_h = [x, x + h]$ denote the set of all numbers z such that $x \le z \le x + h$. Let x be in (s, t) and let h be such that the interval $I_h = [x, x + h]$ is contained in (s, t). Choose v_h in I_h such that $f(v_h) h = F(x + h) - F(x)$. [1]

---

[1]This is the tricky part. In Figure 2.1.4. Take $x = 1$ and $h = 2$ so that $I_2 = [1, 3]$. The signed area between I_h and the graph of f is $F(x + h) - F(x) = 1.3 + 1.7 = 3$. If we take $v_h = 2$, then $f(v_h) = 1.5$ and $f(v_h) h = 1.5 \times 2 = 3 = F(x + h) - F(x)$. Note $1 = m_h \le f(v_h) \le M_h = 1.6$ where m_h is the minimum of f on I_h and M_h is the maximum. Try some other I_h in the figure (e.g.,$[-2, -1]$ or $[-1, +1]$).

Let m_h be the minimum value of f on I_h and let M_h be the maximum. Clearly, $m_h \leq f(v_h) \leq M_h$. As $h > 0$ tends to zero, m_h and M_h tend to $f(x)$ by the continuity of f. This means that $f(v_h)$, which is between m_h and M_h, tends to $f(x)$ also. Thus, $f(v_h)\, h = F(x + h) - F(x)$ implies that

$$\lim_{h \to 0} \frac{F(x + h) - F(x)}{h} = \lim_{h \to 0} \frac{f(v_h)\, h}{h} = \lim_{h \to 0} f(v_h) = f(x).$$

Thus, the right-hand derivative of $F(x)$ is $f(x)$. Using $I_h = [x - h, x]$ for $h > 0$, the same argument works and shows that the left-hand derivative of $F(x)$ is $f(x)$. Hence, the derivative of $F(x)$ is $f(x)$. $\qquad\square$

(2.1.4) **Figure : Graph of Signed Area Function $F(x) = S_a^x(f)$**

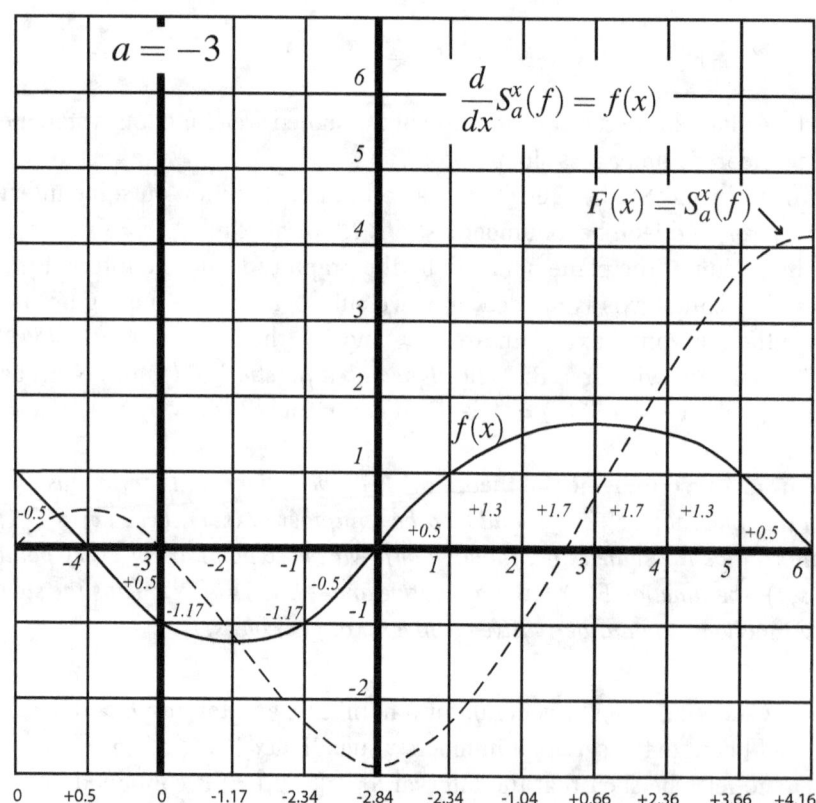

Tutoring points (signed area):

1. Given a function $f(x)$ that is continuous on an interval $[s, t]$, it is easy, at least conceptually, to construct the graph of f. From the graph of f, the construction of a signed area function, and hence an antiderivative F of f, is easy to imagine. For some students, this graphical route to antiderivatives demystifies integral calculus – all to the good.

2. Your students should understand that if $H(x)$ is an integral of $f(x)$ on (s, t), $H(x)$ may not be a signed area function, S_a^x, for any a, $s < a < t$. Why? Find an $H(x)$ that is not zero for every a, $s < a < t$ (recall that S_a^x is always zero at $x = a$). Ask your students to find a specific example of such an $H(x)$. Under the assumptions of Theorem 2.1.3, is it always possible to find an $H(x)$ that is not a signed area function? Draw some examples and use your intuition.

3. Suppose that $H(x)$ is an integral of $f(x)$ on (s, t) and $s < a < t$. Then $F(x) = H(x) - H(a) = S_a^x$. Why? Also note that if a, b, and c are in the interval (s, t) then

$$S_a^b + S_b^c = (H(b) - H(a)) + (H(c) - H(b)) = H(c) - H(a) = S_a^c.$$

Ask your students to explain the formula, $S_a^b + S_b^c = S_a^c$, directly (and graphically) in terms of the definition of the signed area function (Definition 2.1.2).

The Envelope Game

Imagine that you have an envelope and inside is the signed area function of $f(x) = x^2$ with base point $a = 0$. What might you see when you open the envelope? One possibility is that you will see a drawing of a graph of $F(x) = S_0^x(f)$, constructed in a manner analogous to the graph $S_a^x(f)$ shown in Figure 2.1.4. If so, you will know that whoever put such a thing in the envelope wasn't thinking very hard! We learned at the very beginning of this chapter that if $F(x)$ and $G(x)$ are two antiderivatives for $f(x)$ then they differ by a constant function: $F(x) = G(x) + C$. Take $G(x) = x^3/3$, an obvious antiderivative of $f(x) = x^2$. By Theorem 2.1.3, $F(x) = S_0^x(f)$ is also an antiderivative of $f(x)$. Thus $F(x) = x^3/3 + C$ and, since $F(0) = 0$, $C = 0$. What you should find in the envelope is simply

$$S_0^x(f) = x^3/3.$$

The tedious task of computing signed areas, $S_a^x(f)$, of functions $f(x)$ can be replaced by the seemingly very different task of finding concise expressions (e.g., polynomials, rational functions, trigonometric functions) for antiderivatives $F(x)$ of $f(x)$. Computing areas might, at first glance, seem to be too specialized a task – hardly worth developing the powerful techniques of integral calculus. However, many problems in physics, chemistry, engineering, astronomy, etc., present the same basic challenge as computing areas and can be solved using calculus in an analogous manner.

2.2 Definite integrals and Riemann sums

Here is a quick summary: Given a function $f(x)$, any function $G(x)$ such that $\frac{d}{dx}G(x) = f(x)$ is called an antiderivative or integral of $f(x)$. Any two antiderivatives $H(x)$ and $G(x)$ of $f(x)$ satisfy $H(x) - G(x) = C$. Thus, if $H(x)$ and $G(x)$ are antiderivatives of f and $H(a) = G(a)$ for some number a then $C = 0$ and $H(x) = G(x)$ for all values of x.

The signed area function $S_a^x(f)$ is an an antiderivative for f. If an antiderivative $H(x)$ for $f(x)$ satisfies $H(a) = 0$ then, in fact, $H(x) = S_a^x(f)$ (take $G(x) = S_a^x(f)$ and note that $H(a) = G(a) = S_a^a(f) = 0$).

To find the signed area function $S_a^x(f)$, find any antiderivative $F(x)$ of $f(x)$ and set $H(x) = F(x) - F(a)$. Then $H(x) = S_a^x(f)$.

(2.2.1)   **Definite integrals**

The common notation for the signed area function $S_a^x(f)$ in calculus is $\int_a^x f$, or $\int_a^x f(t)dt$, where t is called the *variable of integration*.

It doesn't make any difference what we call the variable of integration:

$$\int_a^x f(t)dt = \int_a^x f(y)dy = \int_a^x f(z)dz = \cdots = S_a^x(f).$$

If we evaluate a signed area function $S_a^x(f)$ of f at a particular number, $x = b$, we designate the answer (a number) by $\int_a^b f(t)dt$. This number is referred to in most calculus books as *the definite integral of f from a to b*. From our previous discussion, one way to compute $\int_a^b f(t)dt$ is to find any antiderivative $F(t)$ of $f(t)$ and compute $F(b) - F(a)$. Thus, $F(b) - F(a) = \int_a^b f(t)dt$, and if we interchange a and b, $\int_b^a f(t)dt = F(a) - F(b)$. Thus,

$$\int_a^b f(t)dt = -\int_b^a f(t)dt.$$

Numbers such as a and b or 0 and 1 that appear at the top and bottom of the integral sign ($\int_a^b$ or $\int_0^1$) are called the *limits of integration*. If the integral sign has no limits, such as $\int f(x)\,dx$, then this stands for just any old antiderivative or integral of f. Sometimes the phrase "indefinite integral $\int f$" is used to mean "integral of f" or "antiderivative of f."

(2.2.2)   **Riemann sums**

Consider Figure 2.2.4 where we see the graph of a function f. We are interested in approximating the definite integral (signed area) from a to b ($\int_a^b f(t)\,dt$).

To approximate $\int_a^b f(t)\,dt$, we have divided the interval $[a, b]$, $a = x_1$, $b = x_9$, into subintervals $[x_1, x_2]$, $[x_2, x_3]$, ..., $[x_8, x_9]$. In each subinterval $[x_i, x_{i+1}]$, we have chosen a number t_i. We then approximate the definite integral, $\int_a^b f(t)\,dt$, by the sum, $\sum_{i=1}^8 f(t_i)(x_{i+1} - x_i)$. To indicate this approximation we use the following notation:

$$\int_a^b f(t)\,dt \approx \sum_{i=1}^8 f(t_i)(x_{i+1} - x_i)\,.$$

If we let $\triangle_i = x_{i+1} - x_i$, this approximation can be written

$$\int_a^b f(t)\,dt \approx \sum_{i=1}^8 f(t_i)\triangle_i\,.$$

Sums such as those used above to approximate definite integrals are called *Riemann sums*.

Definition 2.2.3 (Riemann sum). Let f be a continuous function defined on the interval $[a, b]$. Let $a = x_1 < x_2 < \cdots < x_{n+1} = b$ be points in the interval $[a, b]$, $a < b$. Define $\triangle_i = x_{i+1} - x_i$. Choose t_i in $[x_i, x_{i+1}]$, $i = 1, \ldots, n$. The following sum is called the Riemann sum for f based on the points x_i and t_i:

$$\sum_{i=1}^n f(t_i)\triangle_i\,.$$

29

(2.2.4)   **Figure : Riemann sums**

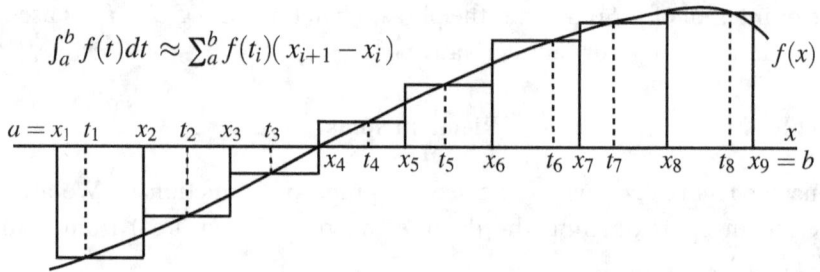

$$\int_a^b f(t)dt \approx \sum_a^b f(t_i)(x_{i+1} - x_i)$$

As we let n get larger and larger in such a way that the $\triangle_i$ goes to zero for all i, the Riemann sum becomes a better and better approximation to the signed area $\int_a^b f(t)dt$. We won't define these limits precisely. In appropriately poorly defined notation, we write

$$\lim \sum_{i=1}^n f(t_i)\triangle_i = \int_a^b f(t)dt .$$

Using the notation of Definition 2.2.3, we can write

$$\sum_{i=1}^n f(t_i)\triangle_i \approx \int_a^b f(t)dt$$

to indicate that we are regarding the Riemann sum as an approximation to the corresponding definite integral. Note that the terms $f(t_i)\triangle_i$ of the Riemann sum are numbers that commute under addition. Thus, $\sum_{i=1}^n f(t_i)\triangle_i = \sum_{i=n}^1 f(t_i)\triangle_i$, and the latter sum should not be interpreted as an approximation for $\int_b^a f(t)dt$. From the definition of the signed area function

$$\int_a^b f(t)dt = -\int_b^a f(t)dt .$$

To approximate $\int_b^a f(t)dt$ where $b > a$, approximate first $\int_a^b f(t)dt$, as in Definition 2.2.3, and change the sign. This is equivalent to changing $\triangle_i$ to $-\triangle_i$, $i = 1, 2, \ldots, n$, in the Riemann sum.

We now work some exercises (Exercises 2.3) using the relationship between integrals and signed areas. Solutions to Exercises 2.3 follow the list of problems. After working these five problems, we suggest you make some minor change to each problem and try to work the problem again. Remember, changes that seem "minor" can sometimes make a problem much more difficult, something you should always be aware of when tutoring students.

30

2.3 Exercises

In these exercises "area" means actual area, not signed area. In Exercise 2.3.1, for example, instead of asking for the area bounded by, "... $x = -1$, $x = 2$, and the horizontal axis," we replaced $x = -1$ by $x = a$ and $x = 2$ by $x = b$. The answer is then expressed in terms of the "parameters" a and b. Calculus gives us the power to solve parameterized problems. You must be able to create such parameterized problems for your students. This exercise will give you experience in making up such problems.

Exercise 2.3.1. Find the area bounded by the graph of $y = x^3$, the lines $x = a$, $x = b$, and the horizontal axis.

Exercise 2.3.2. Find the area bounded by the graph of $y = x^3$, the lines $y = b^3$, $y = a^3$, and the vertical axis.

Exercise 2.3.3. Find the area of the bounded region between the curves $y = -x^2 + c$ and $y = k^x$. Assume $k > 1$ and $c > 1$.

Exercise 2.3.4. Find the area enclosed by the curve $\{(c + a\cos(t), d + b\sin(t)) : 0 \le t < 2\pi\}$.

Exercise 2.3.5. Using polar coordinates, find the area enclosed by the curve $r(\phi) = a + b\cos(\phi)$, $0 \le \phi < 2\pi$, $a > 0$ and $b > 0$.

Solutions to Exercises 2.3

Solution 2.3.1: Assume $a \le b$. This assumption is just a matter of notation and can be made "without loss of generality" (WLOG). The function $f(x) = x^3$ is zero at $x = 0$, negative for $x < 0$, and positive for $x > 0$. A sketch of the situation is shown in Figure 2.3.6 where we show the case $a < 0 < b$.

(2.3.6)   **Figure : Areas and f(x) = x³**

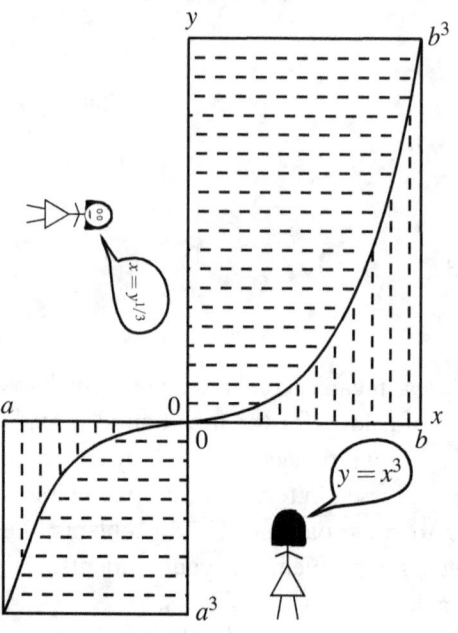

When $a < 0 < b$, the definite integral

$$\int_a^b x^3 dx = \frac{x^4}{4}\Bigg|_a^b = b^4/4 - a^4/4$$

gives the *signed area* between the graph of $f(x) = x^3$ and the interval $[a, b]$ which *does not equal* the area ("actual area"). If $a = -1$ and $b = 2$ the signed area is $2^4/4 - (-1)^4/4 = 15/4$. The area, on the other hand, is given by $\left|\int_a^0 x^3 dx\right| + \left|\int_0^b x^3 dx\right| = a^4/4 + b^4/4$ $(a < 0 < b)$. For $a = -1$ and $b = 2$ the area is $(-1)^4/4 + 2^4/4 = 17/4$. As long as the function $f(x)$ doesn't change sign over an interval, such as the interval $[a, 0]$, then $\left|\int_a^0 f(x)\, dx\right|$ gives the area. If we drop the absolute value, $\int_a^0 x^3 dx$ is the signed area based at $a < 0$ and is a negative number. The sum $\int_0^a x^3 dx + \int_0^b x^3 dx$ expresses the area without absolute value signs (for $a < 0 < b$). If $a \le b \le 0$ or $0 \le a \le b$, the function $f(x) = x^3$ does not change sign over the interval $[a, b]$ so $\left|\int_a^b x^3 dx\right|$ gives the area in both cases.

Solution 2.3.2: Assume $a \le b$ or, equivalently, $a^3 \le b^3$. A sketch of the situation is shown in Figure 2.3.6 where we show the case $a^3 < 0 < b^3$. The

32

area bounded by the graph of $y = x^3$, the lines $y = b^3$, $y = a^3$, and the vertical axis is shown by horizontal dashed lines. Let's take the point of view of the young lady facing us in Figure 2.3.6, namely $x = y^{1/3}$. Reasoning as in Solution 2.3.1, if $a^3 < 0 < b^3$ then the area between the two curves is

$$\int_0^{a^3} y^{1/3}\, dy + \int_0^{b^3} y^{1/3}\, dy = \frac{3}{4} y^{4/3}\Big|_0^{a^3} + \frac{3}{4} y^{4/3}\Big|_0^{b^3} = \frac{3}{4}a^4 + \frac{3}{4}b^4.$$

Alternatively, we can subtract the areas computed in Solution 2.3.1 from the areas of the $a \times a^3$ rectangle and the $b \times b^3$ rectangle:

$$(a^4 - \int_0^a x^3\, dx) + (b^4 - \int_0^b x^3\, dx) = (a^4 - \frac{a^4}{4}) + (b^4 - \frac{b^4}{4}) = \frac{3}{4}a^4 + \frac{3}{4}b^4.$$

As in Solution 2.3.2, if $a \le b \le 0$ or $0 \le a \le b$, the function $x = y^{1/3}$ does not change sign over the interval $[a^3, b^3]$. Thus, the area is $\left|\int_{a^3}^{b^3} y^{1/3}\, dy\right| = \left|\frac{3}{4}b^4 - \frac{3}{4}a^4\right|$.

(2.3.7)   **Figure : Area between y = −x² + c and y = k^x**

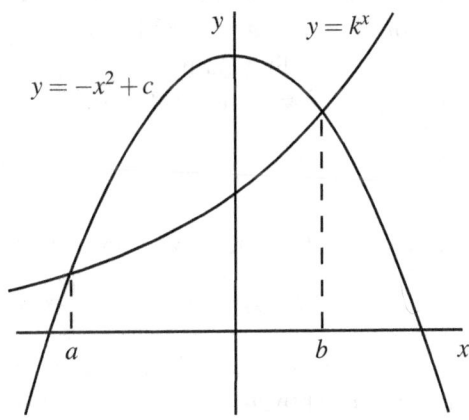

(2.3.8) **Figure : Area bounded by ellipse**

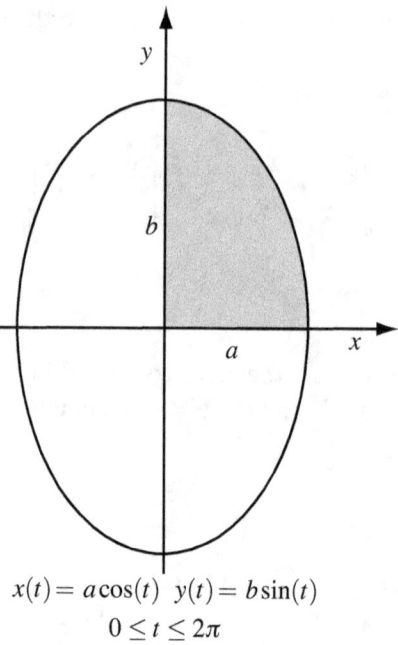

$$x(t) = a\cos(t) \quad y(t) = b\sin(t)$$
$$0 \leq t \leq 2\pi$$

Solution 2.3.3: A sketch of the situation is shown in Figure 2.3.7. The area of the bounded region between the curves $y = -x^2 + c$ and $y = k^x$ ($k > 1$ and $c > 1$) is given by

$$\int_a^b (-x^2 + c - k^x)\, dx = \frac{-x^3}{3} + cx - \left.\frac{k^x}{\ln(k)}\right|_a^b = \frac{a^3 - b^3}{3} + c(b - a) + \frac{k^a - k^b}{\ln(k)}.$$

This expression shows one advantage of an analytical as opposed to graphical or numerical solution. The analytic solution can often be expressed explicitly in terms of parameters (in this case, a, b, c, and k). For a specific k and c, we can solve for a and b (using an online math program, graphing calculator, or math software). For example, if $k = 2$ and $c = 2$ then $a = -1.26$ and $b = +0.65$ (approximately).

Solution 2.3.4: You are asked to find the area enclosed by the curve

$$\{(c + a\cos(t), d + b\sin(t)) : 0 \leq t < 2\pi\}.$$

This curve is an ellipse centered at (c, d). The area is unchanged if we translate this ellipse to one centered at the origin: $\{(a\cos(t), b\sin(t)) : 0 \leq t < 2\pi\}$

(see Figure 2.3.8). By symmetry, we can compute the area of the shaded region in Figure 2.3.8, $\{(a\cos(t), b\sin(t)) : 0 \leq t < \pi/2\}$, and multiply by four. The area of the shaded region is

$$\int_0^a y\,dx = \int_{t=\pi/2}^{t=0} y(t)x'(t)\,dt = ab \int_{\pi/2}^0 \sin(t)(-\sin(t))\,dt = ab \int_0^{\pi/2} \sin^2(t)\,dt.$$

By using trig identities or looking up the integral, the area of the ellipse is

$$4ab \int_0^{\pi/2} \sin^2(t)\,dt = 4ab \left(\frac{t}{2} - \frac{\sin(2t)}{4}\right)\Big|_0^{\pi/2} = 4ab\frac{\pi}{4} = \pi ab.$$

(2.3.9)   **Figure : Area $r(\phi) = a + b\cos(\phi)$, $0 \leq \phi < 2\pi$, $a > 0$, $b > 0$**

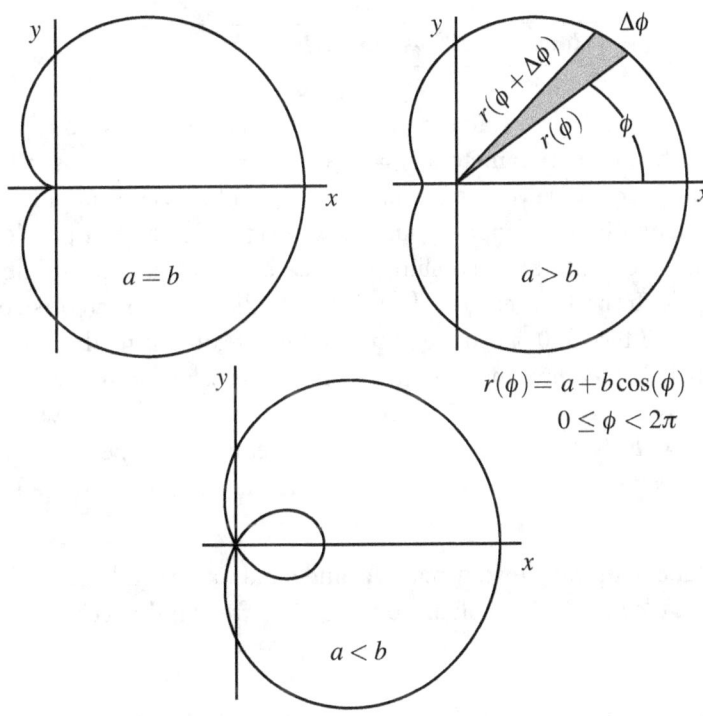

Solution 2.3.5: Using polar coordinates, find the area enclosed by the curve $r(\phi) = a + b\cos(\phi)$, $0 \leq \phi < 2\pi$, $a > 0$ and $b > 0$. Figure 2.3.9 shows the three cases, $a = b$, $a > b$, and $a < b,$. We discuss the first two cases and leave the third case as an additional exercise for the student. The shaded area in Figure 2.3.9 is approximately a triangle with base $r(\phi)\Delta\phi$ and height $r(\phi)$. Thus, its area is approximately $\frac{1}{2}r^2(\phi)\,\Delta\phi$. Reasoning as in Figure 2.1.1, let $A(\phi)$ be the area bounded by the curve r, the x axis, and the ray ϕ. The shaded area is thus $A(\phi + \Delta\phi) - A(\phi) \approx \frac{1}{2}r^2(\phi)\,\Delta\phi$. Taking limits, we write $dA = \frac{1}{2}r^2(\phi)\,d\phi$. Thus, the following integral represents the area of the upper half of the figure (in the cases $a > b$ and $a = b$):

$$A(\pi) = \int_0^\pi \frac{1}{2}r^2(\phi)\,d\phi = \int_0^\pi \frac{1}{2}(a + b\cos(\phi))^2\,d\phi.$$

Expanding the integral on the right, we obtain

$$\int_0^\pi \frac{1}{2}a^2\,d\phi + \int_0^\pi ab\cos(\phi)\,d\phi + \int_0^\pi \frac{1}{2}b^2\cos^2(\phi)\,d\phi.$$

The middle integral is zero. An antiderivative of $\cos^2(\phi)$ is $\phi/2 + \sin(2\phi)/4$. Thus, $A(\pi) = (1/2)\pi a^2 + (1/4)\pi b^2$ and the area enclosed by the curve is $2A(\pi) = \pi a^2 + (1/2)\pi b^2$ (see Figure 2.3.9, cases, $a = b$, $a > b$).

How well did we do with Exercises 2.3?

How well did we do in parametrizing Exercises 2.3? The first four problems could have had more parameters: $y = (x + c)^3$, $y = k^{px+q}$, $c + a\cos(kt)$. These extra parameters, however, would have increased the technical difficulty for the students without providing much new insight. In the last problem, Exercise 2.3.5, we could have combined the cases $a = b$ and $a > b$ into one case, $a \geq b$. In both cases, $a = b$ and $a > b$, the function $r(\phi)$ is continuous over the interval $[0, \pi]$ so the same method of computing the area works. Also, rather than computing the integral from 0 to π and doubling, we could have directly integrated from 0 to 2π. Our most serious mistake was leaving the case $a < b$ "to the student." This is just being lazy, especially since the case $a < b$ requires clarification as to what is meant and justifies a separate discussion.

We conclude with some review material and tables of integrals (from the book *Top-down Calculus*, S. Gill Williamson, available free on the web).

2.4   Inverse trig functions, tables and index

This section contains a Table of Integrals, a table of Trigonometric Functions and an overall Index. The tables are from the book, *Top-down Calculus*, by S. Gill Williamson (with the author's permission).

Differentiation rules for the commonly occurring function-inverse pairs of calculus are a source of difficulty for both students and tutors. We review the basic facts that you should know in a series of figures:

First we review the derivative of the functional inverse of $\sin(x)$ which is denoted by $\sin^{-1}(x)$ or, alternatively, $\arcsin(x)$. We refer to Figure 2.4.1. You will recognize the first graph as that of $y = \sin(x)$ for $-\frac{\pi}{2} \le x \le +\frac{\pi}{2}$. The second graph is that or $y = \arcsin(x)$ for $-1 \le x \le +1$. These functions are functional inverses in that $\sin(\arcsin(x)) = \arcsin(\sin(x)) = x$. The pattern for deriving the derivative $\frac{d}{dx}\arcsin(x) \equiv (\arcsin(x))'$ is, starting with $x = \sin(\arcsin(x))$, differentiate both sides using the chain rule to get $1 = \cos(\arcsin(x))(\arcsin(x))'$. Taking note of the small triangle in Figure 2.4.1, observe that $\arcsin(x) = \arccos((1 - x^2)^{1/2}$ and thus

$$1 = \cos(\arcsin(x))(\arcsin(x))' =$$

$$\cos[\arccos((1 - x^2)^{1/2})](\arcsin(x))' = (1 - x^2)^{1/2}(\arcsin(x))'.$$

Thus,

$$(\arcsin(x))' = \frac{1}{(1 - x^2)^{1/2}}.$$

(2.4.1)

Figure : sin and arcsin

$$y = \arcsin(x) = \sin^{-1}(x)$$

$\sin(x)$

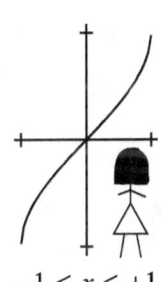
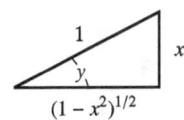

$-\frac{\pi}{2} \le x \le +\frac{\pi}{2}$

$-1 \le x \le +1$

$x = \sin(y) = \sin(\arcsin(x))$

$1 = \cos[\arcsin(x)](\arcsin(x))'$

$\arcsin(x) = \arccos((1 - x^2)^{1/2})$

$1 = ((1 - x^2)^{1/2})(\arcsin(x))'$

$(\arcsin(x))' = \frac{1}{(1 - x^2)^{1/2}}$

(2.4.2)

Figure : cos and arccos

$$y = \arccos(x)$$

$\cos(x)$

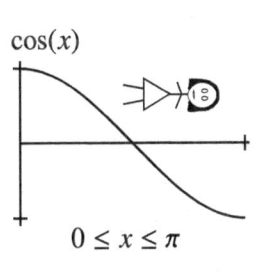
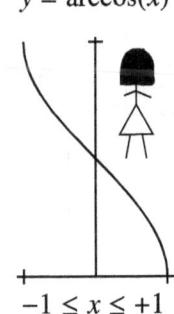

$0 \le x \le \pi$

$-1 \le x \le +1$

$x = \cos(y) = \cos(\arccos(x))$

$1 = -\sin[\arccos(x)](\arccos(x))'$

$\arccos(x) = \arcsin((1 - x^2)^{1/2})$

$(\arccos(x))' = \frac{-1}{(1 - x^2)^{1/2}}$

(2.4.3) **Figure : tan and arctan**

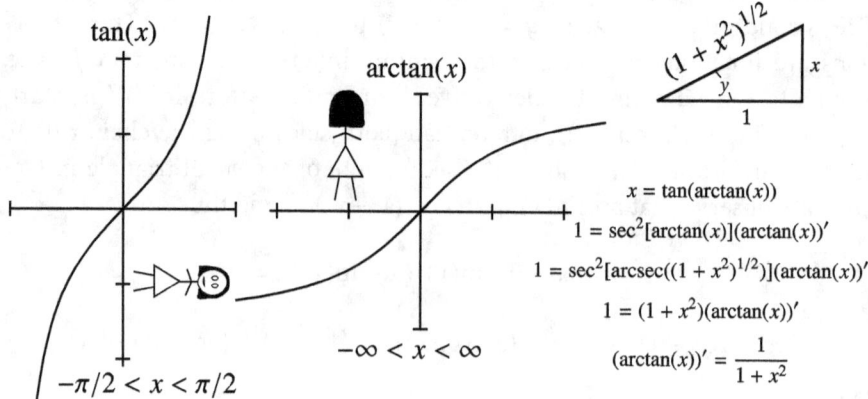

$x = \tan(\arctan(x))$

$1 = \sec^2[\arctan(x)](\arctan(x))'$

$1 = \sec^2[\text{arcsec}((1 + x^2)^{1/2})](\arctan(x))'$

$1 = (1 + x^2)(\arctan(x))'$

$(\arctan(x))' = \dfrac{1}{1 + x^2}$

(2.4.4) **Figure : csc and arccsc**

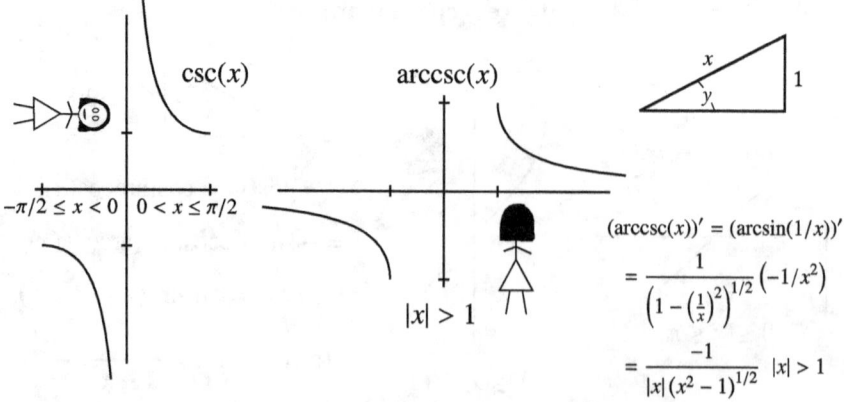

$(\text{arccsc}(x))' = (\arcsin(1/x))'$

$= \dfrac{1}{\left(1 - \left(\frac{1}{x}\right)^2\right)^{1/2}}\left(-1/x^2\right)$

$= \dfrac{-1}{|x|\,(x^2 - 1)^{1/2}} \quad |x| > 1$

(2.4.5)   **Figure : sec and arcsec**

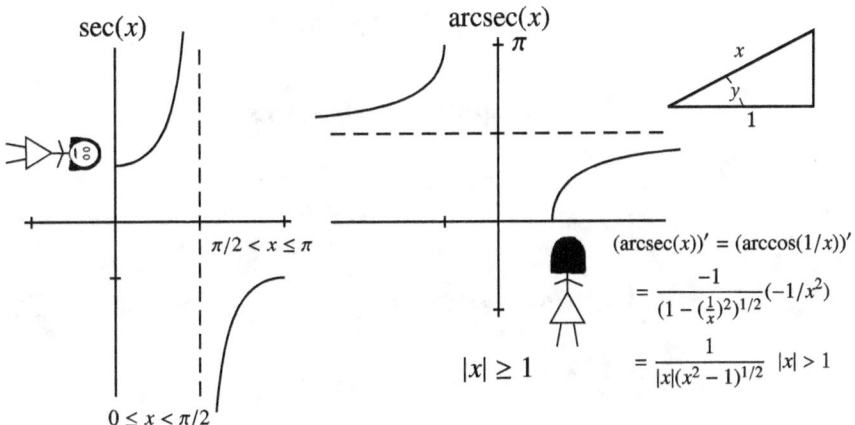

sec(x)

$\pi/2 < x \le \pi$

$0 \le x < \pi/2$

arcsec(x)

π

$|x| \ge 1$

$(\text{arcsec}(x))' = (\text{arccos}(1/x))'$

$= \dfrac{-1}{(1-(\frac{1}{x})^2)^{1/2}}(-1/x^2)$

$= \dfrac{1}{|x|(x^2-1)^{1/2}} \quad |x| > 1$

(2.4.6)

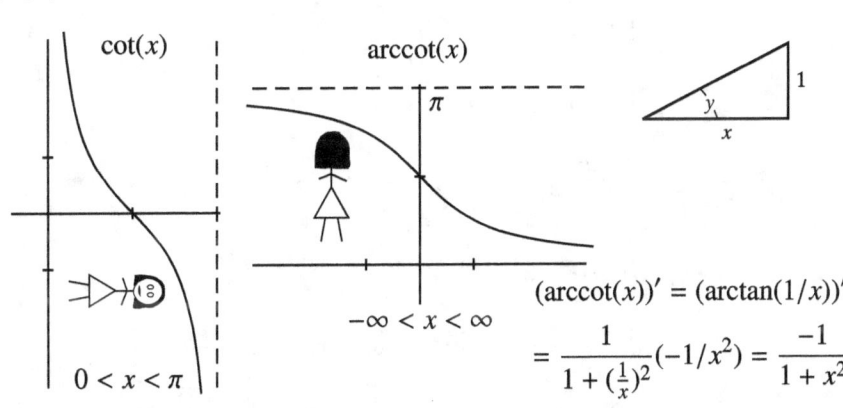

cot(x)

$0 < x < \pi$

arccot(x)

π

$-\infty < x < \infty$

$(\text{arccot}(x))' = (\text{arctan}(1/x))'$

$= \dfrac{1}{1+(\frac{1}{x})^2}(-1/x^2) = \dfrac{-1}{1+x^2}$

TABLE OF INTEGRALS

Fundamental Forms

$$\int a\,dx = ax.$$

$$\int af(x)\,dx = a\int f(x)\,dx.$$

$$\int \frac{dx}{x} = \log x. \quad [\log x = \log(-x) + (2k+1)\pi i.]$$

$$\int x^m\,dx = \frac{x^{m+1}}{m+1}, \text{ when } m \text{ is different from } -1.$$

$$\int e^x\,dx = e^x.$$

$$\int a^x \log a\,dx = a^x.$$

$$\int \frac{dx}{1+x^2} = \tan^{-1}x, \text{ or } -\operatorname{ctn}^{-1}x.$$

$$\int \frac{dx}{\sqrt{1-x^2}} = \sin^{-1}x, \text{ or } -\cos^{-1}x$$

$$\int \frac{dx}{x\sqrt{x^2-1}} = \sec^{-1}x, \text{ or } -\csc^{-1}x.$$

$$\int \frac{dx}{\sqrt{2x-x^2}} = \operatorname{versin}^{-1}x, \text{ or } -\operatorname{coversin}^{-1}x.$$

$$\int \cos x\,dx = \sin x, \text{ or } -\operatorname{coversin} x.$$

$$\int \sin x\,dx = -\cos x, \text{ or } \operatorname{versin} x.$$

$$\int \operatorname{ctn} x\,dx = \log \sin x.$$

$$\int \tan x\,dx = -\log \cos x.$$

$$\int \tan x \sec x\,dx = \sec x.$$

$$\int \sec^2 x\, dx = \tan x.$$

$$\int \csc^2 x\, dx = -\operatorname{ctn} x.$$

In the following formulas, u, v, w, and y represent any functions of x:

$$\int (u + v + w + \text{etc.})\, dx = \int u\, dx + \int v\, dx + \int w\, dx + \text{etc.}$$

$$\int u\, dv = uv - \int v\, du.$$

$$\int u \frac{dv}{dx}\, dx = uv - \int v \frac{du}{dx}\, dx.$$

$$\int f(y)\, dx = \int \frac{f(y)\, dy}{\frac{dy}{dx}}$$

Rational Algebraic Functions

Expressions Involving $(a + bx)$.

The substitution of y or z for x, where $y \equiv a + bx$, $z \equiv (a + bx)/x$, gives

$$\int (a + bx)^m\, dx = \frac{1}{b} \int y^m\, dy.$$

$$\int x\, (a + bx)^m\, dx = \frac{1}{b^2} \int y^m\, (y - a)\, dy.$$

$$\int x^n\, (a + bx)^m\, dx = \frac{1}{b^{n+1}} \int y^m\, (y - a)^n\, dy.$$

$$\int \frac{x^n\, dx}{(a + bx)^m} = \frac{1}{b^{n+1}} \int \frac{(y - a)^n\, dy}{y^m}.$$

$$\int \frac{dx}{x^n\, (a + bx)^m} = -\frac{1}{a^{m+n-1}} \int \frac{(z - b)^{m+n-2}\, dz}{z^m}.$$

Whence

$$\int \frac{dx}{a + bx} = \frac{1}{b} \log\, (a + bx).$$

$$\int \frac{dx}{(a+bx)^2} = -\frac{1}{b(a+bx)}.$$

$$\int \frac{dx}{(a+bx)^3} = -\frac{1}{2b(a+bx)^2}.$$

$$\int \frac{x\,dx}{a+bx} = \frac{1}{b^2}[a+bx - a\log(a+bx)].$$

$$\int \frac{x\,dx}{(a+bx)^2} = \frac{1}{b^2}\left[\log(a+bx) + \frac{a}{a+bx}\right].$$

EXPRESSIONS INVOLVING $(a+bx^n)$.

$$\int \frac{dx}{c^2+x^2} = \frac{1}{c}\tan^{-1}\frac{x}{c} = \frac{1}{c}\sin^{-1}\frac{x}{\sqrt{x^2+c^2}}.$$

$$\int \frac{dx}{c^2-x^2} = \frac{1}{2c}\log\frac{c+x}{c-x}, \quad \int \frac{dx}{x^2-c^2} = \frac{1}{2c}\log\frac{x-c}{x+c}.\ \ ^{*}$$

$$\int \frac{dx}{a+bx^2} = \frac{1}{\sqrt{ab}}\tan^{-1}\left(x\sqrt{\frac{b}{a}}\right),\ \text{or}\ \frac{1}{\sqrt{-ab}}\cdot\tanh^{-1}\left(x\sqrt{\frac{-b}{a}}\right).$$

$$\int \frac{dx}{a+bx^2} = \frac{1}{2\sqrt{-ab}}\log\frac{\sqrt{a}+x\sqrt{-b}}{\sqrt{a}-x\sqrt{-b}},\ \text{if}\ a>0,\ b<0.$$

$$\int \frac{dx}{(a+bx^2)^2} = \frac{x}{2a(a+bx^2)} + \frac{1}{2a}\int \frac{dx}{a+bx^2}.$$

$$\int \frac{dx}{(a+bx^2)^{m+1}} = \frac{1}{2ma}\frac{x}{(a+bx^2)^m} + \frac{2m-1}{2ma}\int \frac{dx}{(a+bx^2)^m}.$$

$$\int \frac{x\,dx}{a+bx^2} = \frac{1}{2b}\log\left(x^2+\frac{a}{b}\right).$$

$$\int \frac{x\,dx}{(a+bx^2)^{m+1}} = \frac{1}{2}\int \frac{dz}{(a+bz)^{m+1}},\ \text{where}\ z = x^2.$$

$$\int \frac{dx}{x(a+bx^2)} = \frac{1}{2a}\log\frac{x^2}{a+bx^2}.$$

$$\int \frac{x^2\,dx}{a+bx^2} = \frac{x}{b} - \frac{a}{b}\int \frac{dx}{a+bx^2}.$$

$$\int \frac{dx}{x^2(a+bx^2)} = -\frac{1}{ax} - \frac{b}{a}\int \frac{dx}{a+bx^2}.$$

$$\int \frac{x^2\,dx}{(a+bx^2)^{m+1}} = \frac{-x}{2\,mb\,(a+bx^2)^m} + \frac{1}{2\,mb}\int \frac{dx}{(a+bx^2)^m}.$$

$$\int \frac{dx}{x^2(a+bx^2)^{m+1}} = \frac{1}{a}\int \frac{dx}{x^2(a+bx^2)^m} - \frac{b}{a}\int \frac{dx}{(a+bx^2)^{m+1}}.$$

$$\sideset{^*}{}{\int} \frac{dx}{c^2-x^2} = \frac{1}{c}\tanh^{-1}\left(\frac{x}{c}\right); \quad \int \frac{dx}{x^2-c^2} = -\frac{1}{c}\operatorname{ctnh}^{-1}\left(\frac{x}{c}\right).$$

Expressions Involving $(a+bx+cx^2)$.

Let $X = a + bx + cx^2$ and $q = 4\,ac - b^2$, then

$$\int \frac{dx}{X} = \frac{2}{\sqrt{q}}\tan^{-1}\frac{2\,cx+b}{\sqrt{q}}, \text{ or } -\frac{2}{\sqrt{-q}}\cdot\tanh^{-1}\frac{2\,cx+b}{\sqrt{-q}}.$$

$$\int \frac{dx}{X} = \frac{1}{\sqrt{-q}}\log\frac{2\,cx+b-\sqrt{-q}}{2\,cx+b+\sqrt{-q}}, \text{ when } q<0.$$

$$\int \frac{dx}{X^2} = \frac{2\,cx+b}{qX} + \frac{2\,c}{q}\int \frac{dx}{X}.$$

$$\int \frac{dx}{X^3} = \frac{2\,cx+b}{q}\left(\frac{1}{2\,X^2} + \frac{3\,c}{qX}\right) + \frac{6\,c^2}{q^2}\int \frac{dx}{X}.$$

$$\int \frac{dx}{X^{n+1}} = \frac{2\,cx+b}{nqX^n} + \frac{2\,(2\,n-1)\,c}{qn}\int \frac{dx}{X^n}.$$

$$\int \frac{x\,dx}{X} = \frac{1}{2\,c}\log X - \frac{b}{2\,c}\int \frac{dx}{X}.$$

$$\int \frac{x\,dx}{X^2} = -\frac{bx+2\,a}{qX} - \frac{b}{q}\int \frac{dx}{X}.$$

$$\int \frac{x\,dx}{X^{n+1}} = -\frac{2\,a+bx}{nqX^n} - \frac{b\,(2\,n-1)}{nq}\int \frac{dx}{X^n}.$$

$$\int \frac{x^2}{X}\,dx = \frac{x}{c} - \frac{b}{2\,c^2}\log X + \frac{b^2-2\,ac}{2\,c^2}\int \frac{dx}{X}.$$

$$\int \frac{x^2}{X^2}\,dx = \frac{(b^2-2\,ac)\,x+ab}{cqX} + \frac{2\,a}{q}\int \frac{dx}{X}.$$

$$\int \frac{x^m\,dx}{X^{n+1}} = -\frac{x^{m-1}}{(2\,n-m+1)\,cX^n} - \frac{n-m+1}{2\,n-m+1}\cdot\frac{b}{c}\int \frac{x^{m-1}\,dx}{X^{n+1}}$$
$$+ \frac{m-1}{2\,n-m+1}\cdot\frac{a}{c}\int \frac{x^{m-2}\,dx}{X^{n+1}}.$$

43

Irrational Algebraic Functions

Expressions Involving $\sqrt{a+bx}$.

The substitution of a new variable of integration, $y = \sqrt{a+bx}$, gives

$$\int \sqrt{a+bx}\, dx = \frac{2}{3\,b}\sqrt{(a+bx)^3}.$$

$$\int x\sqrt{a+bx}\, dx = -\frac{2(2a-3bx)\sqrt{(a+bx)^3}}{15\,b^2}.$$

$$\int x^2\sqrt{a+bx}\, dx = \frac{2(8a^2-12abx+15b^2x^2)\sqrt{(a+bx)^3}}{105\,b^3}$$

$$\int \frac{\sqrt{a+bx}}{x}\, dx = 2\sqrt{a+bx} + a\int \frac{dx}{x\sqrt{a+bx}}.$$

$$\int \frac{dx}{\sqrt{a+bx}} = \frac{2\sqrt{a+bx}}{b}.$$

$$\int \frac{x\,dx}{\sqrt{a+bx}} = -\frac{2(2a-bx)}{3\,b^2}\sqrt{a+bx}.$$

$$\int \frac{x^2\,dx}{\sqrt{a+bx}} = \frac{2(8a^2-4abx+3b^2x^2)}{15\,b^3}\sqrt{a+bx}.$$

$$\int \frac{dx}{x\sqrt{a+bx}} = \frac{1}{\sqrt{a}}\log\left(\frac{\sqrt{a+bx}-\sqrt{a}}{\sqrt{a+bx}+\sqrt{a}}\right), \text{ for } a>0.$$

$$\int \frac{dx}{x\sqrt{a+bx}} = \frac{2}{\sqrt{-a}}\tan^{-1}\sqrt{\frac{a+bx}{-a}}, \text{ or } \frac{-2}{\sqrt{a}}\cdot\tanh^{-1}\sqrt{\frac{a+bx}{a}}.$$

$$\int \frac{x^m\,dx}{\sqrt{a+bx}} = \frac{2\,x^m\sqrt{a+bx}}{(2m+1)b} - \frac{2\,ma}{(2m+1)b}\int \frac{x^{m-1}\,dx}{\sqrt{a+bx}}.$$

$$\int \frac{dx}{x^n\sqrt{a+bx}} = -\frac{\sqrt{a+bx}}{(n-1)ax^{n-1}} - \frac{(2n-3)b}{(2n-2)a}\int \frac{dx}{x^{n-1}\sqrt{a+bx}}.$$

$$\int \sqrt{x^2 \pm a^2}\, dx = \tfrac{1}{2} \left[x \sqrt{x^2 \pm a^2} \pm a^2 \log(x + \sqrt{x^2 \pm a^2}) \right].^{*}$$

$$\int \sqrt{a^2 - x^2}\, dx = \tfrac{1}{2} \left(x \sqrt{a^2 - x^2} + a^2 \sin^{-1}\frac{x}{a} \right).$$

$$\int \frac{dx}{\sqrt{x^2 \pm a^2}} = \log(x + \sqrt{x^2 \pm a^2}).^{*}$$

$$\int \frac{dx}{\sqrt{a^2 - x^2}} = \sin^{-1}\frac{x}{a}, \text{ or } -\cos^{-1}\frac{x}{a}.$$

$$\int \frac{dx}{x\sqrt{x^2 - a^2}} = \frac{1}{a}\cos^{-1}\frac{a}{x}, \text{ or } \frac{1}{u}\sec^{-1}\frac{x}{a}.$$

$$\int \frac{dx}{x\sqrt{a^2 \pm x^2}} = -\frac{1}{a}\log\left(\frac{a + \sqrt{a^2 \pm x^2}}{x} \right)$$

$$\int \frac{\sqrt{a^2 \pm x^2}}{x}\, dx = \sqrt{a^2 \pm x^2} - a \log\frac{a + \sqrt{u^2 \pm x^2}}{x}.^{*}$$

$$\int \frac{\sqrt{x^2 - a^2}}{x}\, dx = \sqrt{x^2 - a^2} - a \cos^{-1}\frac{a}{x}.$$

$$\int \frac{x\, dx}{\sqrt{a^4 \pm x^2}} = \pm \sqrt{a^2 \pm x^2}.$$

$$\int \frac{x\, dx}{\sqrt{x^2 - a^2}} = \sqrt{x^2 - a^2}.$$

$$\int x\sqrt{x^2 \pm a^2}\, dx = \tfrac{1}{8}\sqrt{(x^2 \pm a^2)^3}.$$

$$\int x\sqrt{a^2 - x^2}\, dx = -\tfrac{1}{8}\sqrt{(a^2 - x^2)^3}.$$

$^{*}\log\left(\frac{x + \sqrt{x^2 + a^2}}{a}\right) = \sinh^{-1}\left(\frac{x}{a}\right); \ \log\left(\frac{x + \sqrt{x^2 - a^2}}{a}\right) = \cosh^{-1}\left(\frac{x}{a}\right);$

$\log\left(\frac{a + \sqrt{a^2 - x^2}}{x}\right) = \operatorname{sech}^{-1}\left(\frac{x}{a}\right); \ \log\left(\frac{a + \sqrt{a^2 + x^2}}{x}\right) = \operatorname{csch}^{-1}\left(\frac{x}{a}\right).$

$$\int \sqrt{(x^2 \pm a^2)^3}\, dx$$

$$= \tfrac{1}{4}\left[x\sqrt{(x^2 \pm a^2)^3} \pm \frac{3a^2 x}{2}\sqrt{x^2 \pm a^2} + \frac{3a^4}{2}\log(x + \sqrt{x^2 \pm a^2}) \right].^{*}$$

$$\int \sqrt{(a^2 - x^2)^3}\, dx$$

$$= \tfrac{1}{4}\left[x\sqrt{(a^2 - x^2)^3} + \frac{3\,a^2 x}{2}\sqrt{a^2 - x^2} + \frac{3\,a^4}{2}\sin^{-1}\frac{x}{a} \right].$$

$$\int \frac{dx}{\sqrt{(x^2 \pm a^2)^3}} = \frac{\pm\, x}{a^2\sqrt{x^2 \pm a^2}}.$$

$$\int \frac{dx}{\sqrt{(a^2 - x^2)^3}} = \frac{x}{a^2\sqrt{a^2 - x^2}}.$$

$$\int \frac{x\, dx}{\sqrt{(x^2 \pm a^2)^3}} = \frac{-1}{\sqrt{x^2 \pm a^2}}.$$

$$\int \frac{x\, dx}{\sqrt{(a^2 - x^2)^3}} = \frac{1}{\sqrt{a^2 - x^2}}.$$

$$\int x\sqrt{(x^2 \pm a^2)^3}\, dx = \tfrac{1}{5}\sqrt{(x^2 \pm a^2)^5}.$$

$$\int x\sqrt{(a^2 - x^2)^3}\, dx = -\tfrac{1}{5}\sqrt{(a^2 - x^2)^5}.$$

$$\int x^2\sqrt{x^2 \pm a^2}\, dx$$

$$= \frac{x}{4}\sqrt{(x^2 \pm a^2)^3} \mp \frac{a^2}{8}\,x\sqrt{x^2 \pm a^2} - \frac{a^4}{8}\log\left(x + \sqrt{x^2 \pm a^2}\right).^{*}$$

$$\int x^2\sqrt{a^2 - x^2}\, dx$$

$$= -\frac{x}{4}\sqrt{(a^2 - x^2)^3} + \frac{a^2}{8}\left(x\sqrt{a^2 - x^2} + a^2\sin^{-1}\frac{x}{a} \right).$$

$$\log z = \sinh^{-1}\left(\frac{z^2 - 1}{2\,z}\right) = \cosh^{-1}\left(\frac{z^2 + 1}{2\,z}\right); \ \tanh^{-1} z = -\,i\cdot\tan^{-1}(zi).$$

$$\int \frac{\sqrt{a^2 \pm x^2}\, dx}{x^3} = -\frac{\sqrt{a^2 \pm x^2}}{2\,x^2} \pm \frac{1}{2}\int \frac{dx}{x\sqrt{a^2 \pm x^2}}.$$

$$\int x^3\sqrt{a^2 \pm x^2}\, dx = \left(\pm\tfrac{1}{5}x^2 - \tfrac{2}{15}a^2 \right)\sqrt{(a^2 \pm x^2)^3}.$$

$$\int \frac{dx}{x^3\sqrt{a^2 \pm x^2}} = -\frac{\sqrt{a^2 \pm x^2}}{2\,a^2 x^2} \mp \frac{1}{2\,a^2}\int \frac{dx}{x\sqrt{a^2 \pm x^2}}.$$

$$\int \frac{dx}{x^3 \sqrt{x^2 - a^2}} = \frac{\sqrt{x^2 - a^2}}{2\,a^2 x^2} + \frac{1}{2\,a^3} \sec^{-1}\left(\frac{x}{a}\right).$$

$$\int \frac{x^2\,dx}{\sqrt{x^2 \pm a^2}} = \frac{x}{2}\sqrt{x^2 \pm a^2} \mp \frac{a^2}{2} \log\left(x + \sqrt{x^2 \pm a^2}\right).^{*}$$

$$\int \frac{x^2\,dx}{\sqrt{a^2 - x^2}} = -\frac{x}{2}\sqrt{a^2 - x^2} + \frac{a^2}{2} \sin^{-1}\frac{x}{a}.$$

$$\int \frac{dx}{x^2 \sqrt{x^2 \pm a^2}} = \mp \frac{\sqrt{x^2 \pm a^2}}{a^2 x}.$$

$$\int \frac{dx}{x^2 \sqrt{a^2 - x^2}} = -\frac{\sqrt{a^2 - x^2}}{a^2 x}.$$

$$\int \frac{\sqrt{x^2 \pm a^2}\,dx}{x^2} = -\frac{\sqrt{x^2 \pm a^2}}{x} + \log\left(x + \sqrt{x^2 \pm a^2}\right).^{*}$$

$$\int \frac{\sqrt{a^2 - x^2}}{x^2}\,dx = -\frac{\sqrt{a^2 - x^2}}{x} - \sin^{-1}\frac{x}{a}.$$

$$\int \frac{x^2\,dx}{\sqrt{(x^2 \pm a^2)^3}} = \frac{-x}{\sqrt{x^2 \pm a^2}} + \log\left(x + \sqrt{x^2 \pm a^2}\right).^{*}$$

$$\int \frac{x^2\,dx}{\sqrt{(a^2 - x^2)^3}} = \frac{x}{\sqrt{a^2 - x^2}} - \sin^{-1}\frac{x}{a}.$$

Expressions Involving $\sqrt{a + bx + cx^2}$.

Let $X = a + bx + cx^2$, $q = 4\,ac - b^2$, and $k = \dfrac{4\,c}{q}$. In order to rationalize the function $f(x, \sqrt{a + bx + cx^2})$ we may put $\sqrt{a + bx + cx^2} = \sqrt{\pm c}\,\sqrt{A + Bx \pm x^2}$, according as c is positive or negative, and then substitute for x a new variable z, such that

$$z = \sqrt{A + Bx + x^2} \pm x, \text{ if } c > 0.$$

$$z = \frac{\sqrt{A + Bx - x^2} - \sqrt{A}}{x}, \text{ if } c < 0 \text{ and } \frac{a}{-c} > 0.$$

$$z = \sqrt{\frac{x - \beta}{a - x}}, \text{ where } a \text{ and } \beta \text{ are the roots of the equation}$$

$$A + Bx - x^2 = 0, \text{ if } c < 0 \text{ and } \frac{a}{-c} < 0.$$

47

By rationalization, or by the aid of reduction formulas, may be obtained the values of the following integrals:

$$\int \frac{dx}{\sqrt{X}} = \frac{1}{\sqrt{c}} \log\left(\sqrt{X} + x\sqrt{c} + \frac{b}{2\sqrt{c}}\right), \text{ if } c > 0.$$

$$\int \frac{dx}{\sqrt{X}} = \frac{-1}{\sqrt{-c}} \sin^{-1}\left(\frac{2cx+b}{\sqrt{-q}}\right), \text{ or } \frac{1}{\sqrt{c}} \sinh^{-1}\left(\frac{2cx+b}{\sqrt{q}}\right).$$

$$\int \frac{dx}{x^4 + a^4} = \frac{1}{4a^3\sqrt{2}}\left\{ \log\left(\frac{x^2 + ax\sqrt{2} + a^2}{x^2 - ax\sqrt{2} + a^2}\right) + 2\tan^{-1}\left(\frac{ax\sqrt{2}}{a^2 - x^2}\right)\right\}.$$

$$\int \frac{dx}{x^4 - a^4} = \frac{1}{4a^3}\left\{ \log\left(\frac{x-a}{x+a}\right) - 2\tan^{-1}\left(\frac{x}{a}\right)\right\}.$$

Transcendental Functions

$$\int \sin^2 x\, dx = -\tfrac{1}{2}\cos x \sin x + \tfrac{1}{2}x = \tfrac{1}{2}x - \tfrac{1}{4}\sin 2x.$$

$$\int \sin^3 x\, dx = -\tfrac{1}{8}\cos x\,(\sin^2 x + 2).$$

$$\int \sin^n x\, dx = -\frac{\sin^{n-1} x \cos x}{n} + \frac{n-1}{n}\int \sin^{n-2} x\, dx.$$

$$\int \cos x\, dx = \sin x.$$

$$\int \cos^2 x\, dx = \tfrac{1}{2}\sin x \cos x + \tfrac{1}{2}x = \tfrac{1}{2}x + \tfrac{1}{4}\sin 2x.$$

$$\int \cos^3 x\, dx = \tfrac{1}{8}\sin x\,(\cos^2 x + 2).$$

$$\int \cos^n x\, dx = \frac{1}{n}\cos^{n-1} x \sin x + \frac{n-1}{n}\int \cos^{n-2} x\, dx.$$

$$\int \sin x \cos x\, dx = \tfrac{1}{2}\sin^2 x.$$

$$\int \sin^2 x \cos^2 x\, dx = -\tfrac{1}{8}\left(\tfrac{1}{4}\sin 4x - x\right).$$

$$\int \sin x \cos^m x\, dx = -\frac{\cos^{m+1} x}{m+1}.$$

$$\int \sin^m x \cos x \, dx = \frac{\sin^{m+1} x}{m+1}.$$

$$\int \cos^m x \sin^n x \, dx = \frac{\cos^{m-1} x \sin^{n+1} x}{m+n}$$
$$+ \frac{m-1}{m+n} \int \cos^{m-2} x \sin^n x \, dx.$$

$$\int \cos^m x \sin^n x \, dx = -\frac{\sin^{n-1} x \cos^{m+1} x}{m+n}$$
$$+ \frac{n-1}{m+n} \int \cos^m x \sin^{n-2} x \, dx.$$

$$\int \frac{dx}{\cos^n x} = \frac{1}{n-1} \cdot \frac{\sin x}{\cos^{n-1} x} + \frac{n-2}{n-1} \int \frac{dx}{\cos^{n-2} x}.$$

$$\int \tan x \, dx = -\log \cos x.$$

$$\int \tan^2 x \, dx = \tan x - x.$$

$$\int \tan^n x \, dx = \frac{\tan^{n-1} x}{n-1} - \int \tan^{n-2} x \, dx.$$

$$\int \operatorname{ctn} x \, dx = \log \sin x.$$

$$\int \operatorname{ctn}^2 x \, dx = -\operatorname{ctn} x - x.$$

$$\int \operatorname{ctn}^n x \, dx = -\frac{\operatorname{ctn}^{n-1} x}{n-1} - \int \operatorname{ctn}^{n-2} x \, dx.$$

$$\int \sec x \, dx = \log \tan \left(\frac{\pi}{4} + \frac{x}{2} \right) = \tfrac{1}{2} \log \frac{1 + \sin x}{1 - \sin x}.$$

$$\int \sec^2 x \, dx = \tan x.$$

$$\int \sec^n x \, dx = \int \frac{dx}{\cos^n x} = \frac{\sin x}{(n-1) \cos^{n-1} x} + \frac{n-2}{n-1} \int \frac{dx}{\cos^{n-2} x}$$
$$= \frac{\sin x}{(n-1) \cos^{n-1} x} + \frac{n-2}{n-1} \int \sec^{n-2} x \, dx.$$

$$\int \csc x \, dx = \log \tan \tfrac{1}{2} x.$$

$$\int \csc^n x \, dx = \int \frac{dx}{\sin^n x}$$

$$= -\frac{\cos x}{(n-1)\sin^{n-1}x} + \frac{n-2}{n-1}\int \frac{dx}{\sin^{n-2}x}$$

$$= -\frac{\cos x}{(n-1)\sin^{n-1}x} + \frac{n-2}{n-1}\int \csc^{n-2} x \, dx.$$

$$\int \frac{dx}{1+\sin x} = -\tan \left(\tfrac{1}{4}\pi - \tfrac{1}{2}x\right). \quad [\text{See 241.}]$$

$$\int \frac{dx}{1-\sin x} = \operatorname{ctn} \left(\tfrac{1}{4}\pi - \tfrac{1}{2}x\right) = \tan \left(\tfrac{1}{4}\pi + \tfrac{1}{2}x\right).$$

$$\int \frac{dx}{1+\cos x} = \tan \tfrac{1}{2}x, \text{ or } \csc x - \operatorname{ctn} x.$$

$$\int \frac{dx}{1-\cos x} = -\operatorname{ctn}\tfrac{1}{2}x, \text{ or } -\operatorname{ctn} x - \csc x.$$

$$\int \frac{dx}{a \pm b \sin x} = \frac{2 \sec \theta}{a} \cdot \tan^{-1}\left(\sec \theta \cdot \tan \tfrac{1}{2}x \pm \tan \theta\right),$$

if $a > b$, and $b = a \sin \theta$.

$$\int \frac{dx}{a \pm b \sin x} = \frac{\pm \sec a}{b} \log \frac{\sin \tfrac{1}{2}(a \pm x)}{\cos \tfrac{1}{2}(x \mp a)},$$

if $b > a$, and $a = b \sin a$. [See 241.]

$$\int \frac{dx}{a+b \cos x} = \frac{-1}{\sqrt{a^2-b^2}} \cdot \sin^{-1}\left[\frac{b+a \cos x}{a+b \cos x}\right],$$

$$\text{or } \frac{1}{\sqrt{a^2-b^2}} \sin^{-1}\left[\frac{\sqrt{a^2-b^2}\cdot \sin x}{a+b \cos x}\right],$$

$$\text{or } \frac{2}{\sqrt{a^2-b^2}} \tan^{-1}\left[\sqrt{\frac{a-b}{a+b}} \tan \tfrac{1}{2}x\right],$$

$$\text{or } \frac{1}{\sqrt{a^2-b^2}} \tan^{-1}\left[\frac{\sqrt{a^2-b^2}\cdot \sin x}{b+a \cos x}\right],$$

$$\int x^m \cos x\, dx = x^m \sin x - m \int x^{m-1} \sin x\, dx.$$

$$\int \frac{\sin x}{x^m}\, dx = -\frac{1}{m-1} \cdot \frac{\sin x}{x^{m-1}} + \frac{1}{m-1} \int \frac{\cos x}{x^{m-1}}\, dx.$$

$$\int \frac{\cos x}{x^m}\, dx = -\frac{1}{m-1} \cdot \frac{\cos x}{x^{m-1}} - \frac{1}{m-1} \int \frac{\sin x}{x^{m-1}}\, dx.$$

$$\int \frac{\sin x}{x}\, dx = x - \frac{x^3}{3 \cdot 3!} + \frac{x^5}{5 \cdot 5!} - \frac{x^7}{7 \cdot 7!} + \frac{x^9}{9 \cdot 9!} \cdots.$$

$$\int \frac{\cos x}{x}\, dx = \log x - \frac{x^2}{2 \cdot 2!} + \frac{x^4}{4 \cdot 4!} - \frac{x^6}{6 \cdot 6!} + \frac{x^8}{8 \cdot 8!} \cdots.$$

$$\int \frac{x\, dx}{\sin x} = x + \frac{x^3}{3 \cdot 3!} + \frac{7\, x^5}{3 \cdot 5 \cdot 5!} + \frac{31\, x^7}{3 \cdot 7 \cdot 7!} + \frac{127\, x^9}{3 \cdot 5 \cdot 9!} + \cdots$$

$$\int \frac{x\, dx}{\cos x} = \frac{x^2}{2} + \frac{x^4}{4 \cdot 2!} + \frac{5\, x^6}{6 \cdot 4!} + \frac{61\, x^8}{8 \cdot 6!} + \frac{1385\, x^{10}}{10 \cdot 8!} + \cdots.$$

$$\int \frac{x\, dx}{\sin^2 x} = -x \operatorname{ctn} x + \log \sin x.$$

$$\int \frac{x\, dx}{\cos^2 x} = x \tan x + \log \cos x.$$

$$n^2 \int x^m \sin^n x\, dx$$
$$= x^{m-1} \sin^{n-1} x\, (m \sin x - nx \cos x)$$
$$+ n(n-1) \int x^m \sin^{n-2} x\, dx - m(m-1) \int x^{m-2} \sin^n x\, dx.$$

$$n^2 \int x^m \cos^n x\, dx$$
$$= x^{m-1} \cos^{n-1} x\, (m \cos x + nx \sin x)$$
$$+ n(n-1) \int x^m \cos^{n-2} x\, dx - m(m-1) \int x^{m-2} \cos^n x\, dx.$$

$$\int \frac{\sin^n x\, dx}{\cos^m x} = \frac{1}{n-m} \left(-\frac{\sin^{n-1} x}{\cos^{m-1} x} + (n-1) \int \frac{\sin^{n-2} x\, dx}{\cos^m x} \right)$$
$$= \frac{1}{m-1} \left(\frac{\sin^{n+1} x}{\cos^{m-1} x} - (n-m+2) \int \frac{\sin^n x\, dx}{\cos^{m-2} x} \right)$$
$$= \frac{1}{m-1} \left(\frac{\sin^{n-1} x}{\cos^{m-1} x} - (n-1) \int \frac{\sin^{n-2} x\, dx}{\cos^{m-2} x} \right).$$

$$\int \frac{\cos^m x\, dx}{\sin^n x} = -\frac{\cos^{m+1} x}{(n-1)\sin^{n-1} x} - \frac{m-n+2}{n-1}\int \frac{\cos^m x\, dx}{\sin^{n-2} x}$$

$$= \frac{\cos^{m-1} x}{(m-n)\sin^{n-1} x} + \frac{m-1}{m-n}\int \frac{\cos^{m-2} x\, dx}{\sin^n x}$$

$$= -\frac{1}{n-1}\frac{\cos^{m-1} x}{\sin^{n-1} x} - \frac{m-1}{n-1}\int \frac{\cos^{m-2} x\, dx}{\sin^{n-2} x}.$$

$$\int \frac{\sin^m x\, dx}{\cos^n x} = -\int \frac{\cos^m\left(\frac{\pi}{2}-x\right)d\left(\frac{\pi}{2}-x\right)}{\sin^n\left(\frac{\pi}{2}-x\right)}.$$

$$\int \frac{dx}{\sin x \cos x} = \log \tan x.$$

$$\int \frac{dx}{\cos x \sin^2 x} = \log \tan\left(\frac{\pi}{4}+\frac{x}{2}\right) - \csc x.$$

$$\int \frac{dx}{\sin^m x \cos^n x}$$

$$= \frac{1}{n-1}\cdot\frac{1}{\sin^{m-1} x \cdot \cos^{n-1} x} + \frac{m+n-2}{n-1}\int \frac{dx}{\sin^m x \cdot \cos^{n-2} x}$$

$$= -\frac{1}{m-1}\cdot\frac{1}{\sin^{m-1} x \cdot \cos^{n-1} x} + \frac{m+n-2}{m-1}\int \frac{dx}{\sin^{m-2} x \cdot \cos^n x}.$$

$$\int \frac{dx}{\sin^m x} = -\frac{1}{m-1}\cdot\frac{\cos x}{\sin^{m-1} x} + \frac{m-2}{m-1}\int \frac{dx}{\sin^{m-2} x}.$$

$$\int \frac{x\, dx}{1+\sin x} = -x\tan\tfrac{1}{2}(\tfrac{1}{2}\pi - x) + 2\log\cos\tfrac{1}{2}(\tfrac{1}{2}\pi - x).$$

$$\int \frac{x\, dx}{1-\sin x} = x\operatorname{ctn}\tfrac{1}{2}(\tfrac{1}{2}\pi - x) + 2\log\sin\tfrac{1}{2}(\tfrac{1}{2}\pi - x).$$

$$\int \frac{x\, dx}{1+\cos x} = x\tan\tfrac{1}{2}x + 2\log\cos\tfrac{1}{2}x.$$

$$\int \frac{x\, dx}{1-\cos x} = -x\operatorname{ctn}\tfrac{1}{2}x + 2\log\sin\tfrac{1}{2}x.$$

$$\int \frac{\tan x\, dx}{\sqrt{a+b\tan^2 x}} = \frac{1}{\sqrt{b-a}}\cos^{-1}\left(\frac{\sqrt{b-a}}{\sqrt{b}}\cdot\cos x\right).$$

$$\int \frac{dx}{a+b\tan^2 x} = \frac{1}{a-b}\left[x - \sqrt{\frac{b}{a}}\cdot\tan^{-1}\left(\sqrt{\frac{b}{a}}\cdot\tan x\right)\right].$$

$$\int \frac{\tan x\,dx}{a + b \tan x}$$
$$= \frac{1}{a^2 + b^2}\left\{ bx - a\log(a + b\tan x) + a\log\sec x \right\}.$$

$$\int x \sin x\,dx = \sin x - x\cos x.$$

$$\int x^2 \sin x\,dx = 2x\sin x - (x^2 - 2)\cos x.$$

$$\int x^3 \sin x\,dx = (3x^2 - 6)\sin x - (x^3 - 6x)\cos x.$$

$$\int x^m \sin x\,dx = -x^m\cos x + m\int x^{m-1}\cos x\,dx.$$

$$\int x \cos x\,dx = \cos x + x\sin x.$$

$$\int x^2 \cos x\,dx = 2x\cos x + (x^2 - 2)\sin x.$$

$$\int x^3 \cos x\,dx = (3x^2 - 6)\cos x + (x^3 - 6x)\sin x.$$

$$\int \frac{x^m\,dx}{\sin^n x}$$
$$= \frac{1}{(n-1)(n-2)}\left[-\frac{x^{m-1}(m\sin x + (n-2)x\cos x)}{\sin^{n-1}x} \right.$$
$$\left. + (n-2)^2 \int \frac{x^m\,dx}{\sin^{n-2}x} + m(m-1)\int \frac{x^{m-2}\,dx}{\sin^{n-2}x} \right].$$

$$\int \frac{x^m\,dx}{\cos^n x}$$
$$= \frac{1}{(n-1)(n-2)}\left[-\frac{x^{m-1}(m\cos x - (n-2)x\sin x)}{\cos^{n-1}x} \right.$$
$$\left. + (n-2)^2 \int \frac{x^m\,dx}{\cos^{n-2}x} + m(m-1)\int \frac{x^{m-2}\,dx}{\cos^{n-2}x} \right].$$

$$\int \frac{\sin^n x\,dx}{x^m}$$
$$= \frac{1}{(m-1)(m-2)}\left[-\frac{\sin^{n-1}x((m-2)\sin x + nx\cos x)}{x^{m-1}} \right.$$
$$\left. - n^2 \int \frac{\sin^n x\,dx}{x^{m-2}} + n(n-1)\int \frac{\sin^{n-2}x\,dx}{x^{m-2}} \right].$$

$$\int \frac{\cos^n x\, dx}{x^m}$$

$$= \frac{1}{(m-1)(m-2)} \left[\frac{\cos^{n-1} x\, (nx \cos x - (m-2) \cos x)}{x^{m-1}} \right.$$

$$\left. - n^2 \int \frac{\cos^n x\, dx}{x^{m-2}} + n(n-1) \int \frac{\cos^{n-2} x\, dx}{x^{m-2}} \right].$$

$$\int x^p \sin^m x \, \cos^n x\, dx$$

$$= \frac{1}{(m+n)^2} \left[x^{p-1} \sin^m x \, \cos^{n-1} x\, (p \cos x + (m+n) x \sin x) \right.$$

$$+ (n-1)(m+n) \int x^p \sin^m x \, \cos^{n-2} x\, dx$$

$$- mp \int x^{p-1} \sin^{m-1} x \, \cos^{n-1} x\, dx$$

$$\left. - p(p-1) \int x^{p-2} \sin^m x \, \cos^n x\, dx \right].$$

$$= \frac{1}{(m+n)^2} \left[x^{p-1} \sin^{m-1} x \, \cos^n x\, (p \sin x - (m+n) x \cos x) \right.$$

$$+ (m-1)(m+n) \int x^p \sin^{m-2} x \, \cos^n x\, dx$$

$$+ np \int x^{p-1} \sin^{m-1} x \, \cos^{n-1} x\, dx$$

$$\left. - p(p-1) \int x^{p-2} \sin^m x \, \cos^n x\, dx \right].$$

In this book, we use $\sin^{-1} x$ to denote $\dfrac{1}{\sin x}$. For this part of the table we use the classical notation $\sin^{-1} x = arcsin\,(x)$, $\cos^{-1} x = arccos\,(x)$, etc.

$$\int \sin^{-1} x\, dx = x \sin^{-1} x + \sqrt{1-x^2}.$$

$$\int \cos^{-1} x\, dx = x \cos^{-1} x - \sqrt{1-x^2}.$$

$$\int \tan^{-1} x\, dx = x \tan^{-1} x - \tfrac{1}{2} \log(1+x^2).$$

$$\int \operatorname{ctn}^{-1} x\, dx = x \operatorname{ctn}^{-1} x + \tfrac{1}{2} \log(1+x^2).$$

$$\int \sin mx \sin nx\, dx = \frac{\sin (m-n)\, x}{2\, (m-n)} - \frac{\sin (m+n)\, x}{2\, (m+n)}.$$

$$\int \sin mx \cos nx\, dx = -\frac{\cos (m-n)\, x}{2\, (m-n)} - \frac{\cos (m+n)\, x}{2\, (m+n)}.$$

$$\int \cos mx \cos nx\, dx = \frac{\sin (m-n)\, x}{2\, (m-n)} + \frac{\sin (m+n)\, x}{2\, (m+n)}.$$

$$\int \sin^2 mx\, dx = \frac{1}{2\, m}\, (mx - \sin mx \cos mx).$$

$$\int \cos^2 mx\, dx = \frac{1}{2\, m}\, (mx + \sin mx \cos mx).$$

$$\int \sin mx \cos mx\, dx = -\frac{1}{4\, m}\cos 2\, mx.$$

$$\int \sin nx \sin^m x\, dx = \frac{1}{m+n}\left[-\cos nx \sin^m x \right.$$
$$\left. + m \int \cos (n-1)\, x \cdot \sin^{m-1} x\, dx \right]$$

$$\int \sec^{-1} x\, dx = x \sec^{-1} x - \log (x + \sqrt{x^2 - 1}).$$

$$\int \csc^{-1} x\, dx = x \csc^{-1} x + \log (x + \sqrt{x^2 - 1}).$$

$$\int \operatorname{versin}^{-1} x\, dx = (x-1)\operatorname{versin}^{-1} x + \sqrt{2\, x - x^2}.$$

$$\int (\sin^{-1} x)^2\, dx = x (\sin^{-1} x)^2 - 2\, x + 2\sqrt{1-x^2}\, \sin^{-1} x.$$

$$\int (\cos^{-1} x)^2\, dx = x (\cos^{-1} x)^2 - 2\, x - 2\sqrt{1-x^2}\, \cos^{-1} x.$$

$$\int x \sin^{-1} x\, dx = \tfrac{1}{4}\left[(2\, x^2 - 1)\sin^{-1} x + x\sqrt{1-x^2} \right].$$

$$\int x \cos^{-1} x\, dx = \tfrac{1}{4}\left[(2\, x^2 - 1)\cos^{-1} x - x\sqrt{1-x^2} \right].$$

$$\int x \tan^{-1} x\, dx = \tfrac{1}{2}\left[(x^2 + 1)\tan^{-1} x - x \right].$$

55

$$\int x \operatorname{ctn}^{-1}x\,dx = \tfrac{1}{2}\left[(x^2+1)\operatorname{ctn}^{-1}x + x\right].$$

$$\int x \sec^{-1}x\,dx = \tfrac{1}{2}\left[x^2 \sec^{-1}x - \sqrt{x^2-1}\right].$$

$$\int x \csc^{-1}x\,dx = \tfrac{1}{2}\left[x^2 \csc^{-1}x + \sqrt{x^2-1}\right].$$

$$\int x^n \sin^{-1}x\,dx = \frac{1}{n+1}\left(x^{n+1}\sin^{-1}x - \int \frac{x^{n+1}\,dx}{\sqrt{1-x^2}}\right).$$

$$\int x^n \cos^{-1}x\,dx = \frac{1}{n+1}\left(x^{n+1}\cos^{-1}x + \int \frac{x^{n+1}\,dx}{\sqrt{1-x^2}}\right).$$

$$\int x^n \tan^{-1}x\,dx = \frac{1}{n+1}\left(x^{n+1}\tan^{-1}x - \int \frac{x^{n+1}\,dx}{1+x^2}\right).$$

$$\int x^n \operatorname{ctn}^{-1}x\,dx = \frac{1}{n+1}\left(x^{n+1}\operatorname{ctn}^{-1}x + \int \frac{x^{n+1}\,dx}{1+x^2}\right).$$

$$\int \frac{\sin^{-1}x\,dx}{x^2} = \log\left(\frac{1-\sqrt{1-x^2}}{x}\right) - \frac{\sin^{-1}x}{x}.$$

$$\int \frac{\tan^{-1}x\,dx}{x^2} = \log x - \tfrac{1}{2}\log(1+x^2) - \frac{\tan^{-1}x}{x}.$$

$$\int e^{ax}\,dx = \frac{e^{ax}}{a}. \qquad \int f(e^{ax})\,dx = \int \frac{f(y)\,dy}{ay}, \; y = e^{ax}.$$

$$\int x e^{ax}\,dx = \frac{e^{ax}}{a^2}(ax-1).$$

$$\int x^m e^{ax}\,dx = \frac{x^m e^{ax}}{a} - \frac{m}{a}\int x^{m-1}e^{ax}\,dx.$$

$$\int \frac{e^{ax}}{x^m}\,dx = \frac{1}{m-1}\left[-\frac{e^{ax}}{x^{m-1}} + a\int \frac{e^{ax}\,dx}{x^{m-1}}\right].$$

$$\int a^{bx}\,dx = \frac{a^{bx}}{b\log a}. \qquad \int f(a^{bx})\,dx = \int \frac{f(y)\,dy}{b\cdot\log a\cdot y}, \; y = a^{bx}.$$

$$\int x^n a^x\,dx = \frac{a^x x^n}{\log a} - \frac{na^x x^{n-1}}{(\log a)^2} + \frac{n(n-1)a^x x^{n-2}}{(\log a)^3}\cdots$$
$$\pm \frac{n(n-1)(n-2)\cdots 2.1\,a^x}{(\log a)^{n+1}}.$$

$$\int \frac{a^x dx}{x^n} = \frac{1}{n-1}\left[-\frac{a^x}{x^{n-1}} - \frac{a^x \cdot \log a}{(n-2)x^{n-2}}\right.$$

$$\left.- \frac{a^x \cdot (\log a)^2}{(n-2)(n-3)x^{n-3}} - \cdots + \frac{(\log a)^{n-1}}{(n-2)(n-3)\cdots 2.1}\int \frac{a^x dx}{x}\right].$$

$$\int \frac{a^x dx}{x} = \log x + x \log a + \frac{(x \log a)^2}{2 \cdot 2!} + \frac{(x \log a)^3}{3 \cdot 3!} + \cdots.$$

$$\int \frac{\log x\, dx}{(a+bx)^m}$$

$$= \frac{1}{b(m-1)}\left[-\frac{\log x}{(a+bx)^{m-1}} + \int \frac{dx}{x(a+bx)^{m-1}}\right].$$

$$\int \frac{\log x\, dx}{a+bx} = \frac{1}{b}\log x \cdot \log(a+bx) - \frac{1}{b}\int \frac{\log(a+bx)\, dx}{x}.$$

$$\int (a+bx)\log x\, dx = \frac{(a+bx)^2}{2b}\log x - \frac{a^2 \log x}{2b} - ax - \tfrac{1}{4}bx^2.$$

$$\int \frac{\log x\, dx}{\sqrt{a+bx}}$$

$$= \frac{2}{b}\left[(\log x - 2)\sqrt{a+bx} + \sqrt{a}\,\log(\sqrt{a+bx} + \sqrt{a})\right.$$

$$\left.- \sqrt{a}\,\log(\sqrt{a+bx} - \sqrt{a})\right], \text{ if } a > 0$$

$$= \frac{2}{b}\left[(\log x - 2)\sqrt{a+bx} + 2\sqrt{-a}\,\tan^{-1}\sqrt{\frac{a+bx}{-a}}\right], \text{ if } a < 0.$$

$$\int \sin\log x\, dx = \tfrac{1}{2}x[\sin\log x - \cos\log x].$$

$$\int \cos\log x\, dx = \tfrac{1}{2}x[\sin\log x + \cos\log x].$$

$$\int \frac{(\log x)^n dx}{x} = \frac{(\log x)^{n+1}}{n+1}.$$

$$\int \frac{dx}{\log x} = \log(\log x) + \log x + \frac{(\log x)^2}{2 \cdot 2!} + \frac{(\log x)^3}{3 \cdot 3!} + \cdots.$$

$$\int \frac{dx}{(\log x)^n} = -\frac{x}{(n-1)(\log x)^{n-1}} + \frac{1}{n-1}\int \frac{dx}{(\log x)^{n-1}}.$$

$$\int \frac{x^m dx}{(\log x)^n} = -\frac{x^{m+1}}{(n-1)(\log x)^{n-1}} + \frac{m+1}{n-1}\int \frac{x^m dx}{(\log x)^{n-1}}.$$

$$\int \log x \, dx = x \log x - x.$$

$$\int x^m \log x \, dx = x^{m+1} \left[\frac{\log x}{m+1} - \frac{1}{(m+1)^2} \right].$$

$$\int (\log x)^n \, dx = x (\log x)^n - n \int (\log x)^{n-1} \, dx.$$

$$\int x^m (\log x)^n \, dx = \frac{x^{m+1} (\log x)^n}{m+1} - \frac{n}{m+1} \int x^m (\log x)^{n-1} \, dx.$$

$$\int \frac{dx}{x \log x} = \log (\log x), \quad \text{and} \quad \int \frac{(n-1) \, dx}{x (\log x)^n} = \frac{-1}{(\log x)^{n-1}}.$$

$$\int \log (a^2 + x^2) \, dx = x \cdot \log (a^2 + x^2) - 2x + 2a \cdot \tan^{-1} \left(\frac{x}{a} \right).$$

$$\int \frac{dx}{1 + e^x} = \log \frac{e^x}{1 + e^x}.$$

$$\int \frac{dx}{a + be^{mx}} = \frac{1}{am} [mx - \log (a + be^{mx})].$$

$$\int \frac{dx}{ae^{mx} + be^{-mx}} = \frac{1}{m\sqrt{ab}} \tan^{-1} \left(e^{mx} \sqrt{\frac{a}{b}} \right).$$

$$\int \frac{dx}{\sqrt{a + be^{mx}}} = \frac{1}{m\sqrt{a}} \{ \log (\sqrt{a + be^{mx}} - \sqrt{a})$$

$$- \log (\sqrt{a + b \, e^{mx}} + \sqrt{a}) \}, \quad \text{or} \quad \frac{2}{m\sqrt{-a}} \tan^{-1} \frac{\sqrt{a + be^{mx}}}{\sqrt{-a}}.$$

$$\int \frac{xe^x \, dx}{(1+x)^2} = \frac{e^x}{1+x}, \quad \int x^n \cdot e^{ax^{n+1}} \, dx = \frac{e^{ax^{n+1}}}{a(n+1)}.$$

$$\int e^{ax} \sin px \, dx = \frac{e^{ax} (a \sin px - p \cos px)}{a^2 + p^2}.$$

$$\int e^{ax} \cos px \, dx = \frac{e^{ax} (a \cos px + p \sin px)}{a^2 + p^2}.$$

$$\int e^{ax} \log x \, dx = \frac{e^{ax} \log x}{a} - \frac{1}{a} \int \frac{e^{ax} \, dx}{x}.$$

$$\int e^{ax} \sin^2 x \, dx = \frac{e^{ax}}{4 + a^2} \left(\sin x (a \sin x - 2 \cos x) + \frac{2}{a} \right).$$

$$\int e^{ax} \cos^2 x\, dx = \frac{e^{ax}}{4 + a^2}\left(\cos x\,(2 \sin x + a \cos x) + \frac{2}{a} \right).$$

$$\int e^{ax} \sin^n bx\, dx = \frac{1}{a^2 + n^2 b^2}\bigg((a \sin bx$$

$$- nb \cos bx)\, e^{ax} \sin^{n-1} bx + n\,(n - 1)\, b^2 \int e^{ax} \sin^{n-2} bx \cdot dx \bigg).$$

$$\int e^{ax} \cos^n bx\, dx = \frac{1}{a^2 + n^2 b^2}\bigg((a \cos bx$$

$$+ nb \sin bx)\, e^{ax} \cos^{n-1} bx + n\,(n - 1)\, b^2 \int e^{ax} \cos^{n-2} bx\, dx \bigg).$$

$$\int e^{ax} \tan^n x\, dx$$

$$= \frac{e^{ax} \tan^{n-1} x}{n - 1} - \frac{a}{n - 1}\int e^{ax} \tan^{n-1} x\, dx - \int e^{ax} \tan^{n-2} x\, dx.$$

$$\int e^{ax} \operatorname{ctn}^n x\, dx$$

$$= - \frac{e^{ax} \operatorname{ctn}^{n-1} x}{n - 1} + \frac{a}{n - 1}\int e^{ax} \operatorname{ctn}^{n-1} x\, dx - \int e^{ax} \operatorname{ctn}^{n-2} x\, dx.$$

$$\int \frac{e^{ax}\, dx}{\sin^n x} = - e^{ax}\,\frac{a \sin x + (n - 2) \cos x}{(n - 1)\,(n - 2)\, \sin^{n-1} x}$$

$$+ \frac{a^2 + (n - 2)^2}{(n - 1)\,(n - 2)}\int \frac{e^{ax}\, dx}{\sin^{n-2} x}.$$

$$\int \frac{e^{ax}\, dx}{\cos^n x} = - e^{ax}\,\frac{a \cos x - (n - 2) \sin x}{(n - 1)\,(n - 2)\, \cos^{n-1} x}$$

$$+ \frac{a^2 + (n - 2)^2}{(n - 1)\,(n - 2)}\int \frac{e^{ax}\, dx}{\cos^{n-2} x}.$$

$$\int e^{ax} \sin^m x \cos^n x\, dx$$

$$= \frac{1}{(m + n)^2 + a^2}\Big\{ e^{ax} \sin^{m} x \cos^{n-1} x\,(a \cos x + (m + n) \sin x)$$

$$- ma \int e^{ax} \sin^{m-1} x \cos^{m-1} x\, dx$$

$$+ (n - 1)\,(m + n)\int e^{ax} \sin^m x \cos^{n-2} x\, dx \Big\}$$

$$\int \frac{x^m\,dx}{\log x} = \int \frac{e^{-y}}{y}\,dy, \text{ where } y = -(m+1)\log x.$$

MISCELLANEOUS DEFINITE INTEGRALS

$$\int_0^\infty \frac{a\,dx}{a^2 + x^2} = \frac{\pi}{2}, \text{ if } a > 0;\ 0, \text{ if } a = 0;\ -\frac{\pi}{2}, \text{ if } a < 0.$$

$$\int_0^\infty x^{n-1} e^{-x}\,dx = \int_0^1 \left[\log \frac{1}{x}\right]^{n-1} dx \equiv \Gamma(n).$$

$$\Gamma(z+1) = z \cdot \Gamma(z), \text{ if } z > 0.$$

$$\Gamma(y) \cdot \Gamma(1-y) = \frac{\pi}{\sin \pi y}, \text{ if } 1 > y > 0.\quad \Gamma(2) = \Gamma(1) = 1.$$

$$\Gamma(n+1) = n!, \text{ if } n \text{ is an integer.} \qquad \Gamma(z) = \Pi(z-1).$$

$$\Gamma(\tfrac{1}{2}) = \sqrt{\pi}. \qquad Z(y) = D_y[\log \Gamma(y)]. \quad Z(1) = -0.577216.$$

$$\int_0^1 x^{m-1}(1-x)^{n-1}\,dx = \int_0^\infty \frac{x^{m-1}\,dx}{(1+x)^{m+n}} = \frac{\Gamma(m)\,\Gamma(n)}{\Gamma(m+n)}.$$

$$\int_0^{\frac{\pi}{2}} \sin^n x\,dx = \int_0^{\frac{\pi}{2}} \cos^n x\,dx$$

$$= \frac{1\cdot3\cdot5\cdots(n-1)}{2\cdot4\cdot6\cdots(n)} \cdot \frac{\pi}{2}, \text{ if } n \text{ is an even integer,}$$

$$= \frac{2\cdot4\cdot6\cdots(n-1)}{1\cdot3\cdot5\cdot7\cdots n}, \text{ if } n \text{ is an odd integer,}$$

$$= \tfrac{1}{2}\sqrt{\pi}\,\frac{\Gamma\left(\dfrac{n+1}{2}\right)}{\Gamma\left(\dfrac{n}{2}+1\right)}, \text{ for any value of } n \text{ greater than } -1.$$

$$\int_0^\infty \frac{\sin mx\,dx}{x} = \frac{\pi}{2}, \text{ if } m > 0;\ 0, \text{ if } m = 0;\ -\frac{\pi}{2}, \text{ if } m < 0.$$

$$\int_0^1 \log\left(\frac{1+x}{1-x}\right) \cdot \frac{dx}{x} = \frac{\pi^2}{4}.$$

$$\int_0^1 \frac{\log x\,dx}{\sqrt{1-x^2}} = -\frac{\pi}{2}\log 2.$$

$$\int_0^1 \frac{(x^p - x^q)\,dx}{\log x} = \log \frac{p+1}{q+1}, \text{ if } p+1>0,\, q+1>0.$$

$$\int_0^1 (\log x)^n\,dx = (-1)^n \cdot n!.$$

$$\int_0^1 \left(\log \frac{1}{x}\right)^{\frac{1}{2}}\,dx = \frac{\sqrt{\pi}}{2}.$$

$$\int_0^1 \left(\log \frac{1}{x}\right)^n\,dx = n!.$$

$$\int_0^1 \frac{dx}{\sqrt{\log\left(\frac{1}{x}\right)}} = \sqrt{\pi}.$$

$$\int_0^1 x^m \log\left(\frac{1}{x}\right)^n\,dx = \frac{\Gamma(n+1)}{(m+1)^{n+1}}, \text{ if } m+1>0,\, n+1>0.$$

$$\int_0^\infty \log\left(\frac{e^x+1}{e^x-1}\right)\,dx = \frac{\pi^2}{4}.$$

$$\int_0^{\frac{\pi}{2}} \log \sin x\,dx = \int_0^{\frac{\pi}{2}} \log \cos x\,dx = -\frac{\pi}{2}\cdot \log 2.$$

$$\int_0^{\pi} x \cdot \log \sin x\,dx = -\frac{\pi^2}{2}\log 2.$$

$$\int_0^{\pi} \log(a \pm b \cos x)\,dx = \pi \log\left(\frac{a+\sqrt{a^2-b^2}}{2}\right). \qquad a \geqq b.$$

$$\int_0^\infty \frac{dx}{e^{nx}+e^{-nx}} = \frac{\pi}{4n}.$$

$$\int_0^\infty \frac{x\,dx}{e^{nx}-e^{-nx}} = \frac{\pi^2}{8n^2}.$$

$$\int_0^{\pi i} \sinh(mx)\cdot \sinh(nx)\,dx = \int_0^{\pi i} \cosh(mx)\cdot \cosh(nx)\,dx$$
$$= 0, \text{ if } m \text{ is different from } n.$$

$$\int_0^{\pi i} \cosh^2(mx)\,dx = -\int_0^{\pi i} \sinh^2(mx)\,dx = \frac{\pi i}{2}.$$

$$\int_{-\pi i}^{+\pi i} \sinh(mx)\,dx = 0.$$

$$\int_0^{\pi i} \cosh(mx)\,dx = 0.$$

$$\int_{-\pi i}^{\pi i} \sinh(mx)\cosh(nx)\,dx = 0.$$

$$\int_0^{\pi i} \sinh(mx)\cosh(mx)\,dx = 0.$$

$$\int_0^\infty e^{-ax}\cos mx\,dx = \frac{a}{a^2+m^2}, \text{ if } a>0.$$

$$\int_0^\infty e^{-ax}\sin mx\,dx = \frac{m}{a^2+m^2}, \text{ if } a>0.$$

$$\int_0^\infty e^{-a^2 x^2}\cos bx\,dx = \frac{\sqrt{\pi}\cdot e^{-\frac{b^2}{4a^2}}}{2a}. \qquad\qquad a>0.$$

$$\int_0^1 \frac{\log x}{1-x}\,dx = -\frac{\pi^2}{6}.$$

$$\int_0^1 \frac{\log x}{1+x}\,dx = -\frac{\pi^2}{12}.$$

$$\int_0^1 \frac{\log x}{1-x^2}\,dx = -\frac{\pi^2}{8}.$$

$$\int_0^\infty \frac{\sin x \cdot \cos mx\,dx}{x} = 0, \text{ if } m<-1 \text{ or } m>1;$$
$$\frac{\pi}{4}, \text{ if } m=-1 \text{ or } m=1; \ \frac{\pi}{2}, \text{ if } -1<m<1.$$

$$\int_0^\infty \frac{\sin^2 x\,dx}{x^2} = \frac{\pi}{2}.$$

$$\int_0^\infty \cos(x^2)\,dx = \int_0^\infty \sin(x^2)\,dx = \tfrac{1}{2}\sqrt{\frac{\pi}{2}}.$$

$$\int_0^\pi \sin kx \cdot \sin mx\,dx = \int_0^\pi \cos kx \cdot \cos mx\,dx = 0,$$
if k is different from m.

$$\int_0^\pi \sin^2 mx\,dx = \int_0^\pi \cos^2 mx\,dx = \frac{\pi}{2}.$$

$$\int_0^\infty \frac{\cos mx\,dx}{1+x^2} = \frac{\pi}{2}\cdot e^{-m}. \qquad\qquad m>0.$$

$$\int_0^\infty \frac{\cos x\,dx}{\sqrt{x}} = \int_0^\infty \frac{\sin x\,dx}{\sqrt{x}} = \sqrt{\frac{\pi}{2}}.$$

$$\int_0^\infty e^{-a^2 x^2}\, dx = \frac{1}{2a}\sqrt{\pi}\cdot = \frac{1}{2a}\,\Gamma(\tfrac{1}{2}).$$

$$\int_0^\infty x^n e^{-ax}\, dx = \frac{\Gamma(n+1)}{a^{n+1}} = \frac{n!}{a^{n+1}}\cdot$$

$$\int_0^\infty x^{2n} e^{-ax^2}\, dx = \frac{1\cdot 3\cdot 5\cdots(2n-1)}{2^{n+1} a^n}\sqrt{\frac{\pi}{a}}\cdot$$

$$\int_0^\infty e^{-x^2 - \frac{a^2}{x^2}}\, dx = \frac{e^{-2a}\sqrt{\pi}}{2}\cdot \qquad\qquad a>0.$$

$$\int_0^\infty e^{-nx}\sqrt{x}\, dx = \frac{1}{2n}\sqrt{\frac{\pi}{n}}\cdot$$

$$\int_0^\infty \frac{e^{-nx}}{\sqrt{x}}\, dx = \sqrt{\frac{\pi}{n}}\cdot \qquad\qquad a>0.$$

TRIGONOMETRIC FUNCTIONS

	0°.	30°.	45°.	60°.	90°.	120°.	135°.	150°.	180°.
sin	0	$\frac{1}{2}$	$\frac{1}{2}\sqrt{2}$	$\frac{1}{2}\sqrt{3}$	1	$\frac{1}{2}\sqrt{3}$	$\frac{1}{2}\sqrt{2}$	$\frac{1}{2}$	0
cos	1	$\frac{1}{2}\sqrt{3}$	$\frac{1}{2}\sqrt{2}$	$\frac{1}{2}$	0	$-\frac{1}{2}$	$-\frac{1}{2}\sqrt{2}$	$-\frac{1}{2}\sqrt{3}$	-1
tan	0	$\frac{1}{\sqrt{3}}$	1	$\sqrt{3}$	∞	$-\sqrt{3}$	-1	$-\frac{1}{\sqrt{3}}$	0
ctn	∞	$\sqrt{3}$	1	$\frac{1}{\sqrt{3}}$	0	$-\frac{1}{\sqrt{3}}$	-1	$-\sqrt{3}$	∞
sec	1	$\frac{2}{\sqrt{3}}$	$\sqrt{2}$	2	∞	-2	$-\sqrt{2}$	$-\frac{2}{\sqrt{3}}$	-1
cosec	∞	2	$\sqrt{2}$	$\frac{2}{\sqrt{3}}$	1	$\frac{2}{\sqrt{3}}$	$\sqrt{2}$	2	∞

$$\sin \tfrac{1}{2} a = \sqrt{\tfrac{1}{2}(1-\cos a)}.$$

$$\cos \tfrac{1}{2} a = \sqrt{\tfrac{1}{2}(1+\cos a)}.$$

$$\tan \tfrac{1}{2} a = \sqrt{\frac{1-\cos a}{1+\cos a}} = \frac{1-\cos a}{\sin a} = \frac{\sin a}{1+\cos a}\cdot$$

$$\sin 2a = 2\sin a \cos a.$$

$$\sin 3a = 3\sin a - 4\sin^3 a.$$

$$\sin 4a = 8\cos^3 a\cdot\sin a - 4\cos a \sin a.$$

$$\sin 5\,a = 5 \sin a - 20 \sin^3 a + 16 \sin^5 a.$$

$$\sin 6\,a = 32 \cos^5 a \sin a - 32 \cos^3 a \sin a + 6 \cos a \sin a.$$

$$\cos 2\,a = \cos^2 a - \sin^2 a = 1 - 2 \sin^2 a = 2 \cos^2 a - 1.$$

$$\cos 3\,a = 4 \cos^3 a - 3 \cos a.$$

$$\cos 4\,a = 8 \cos^4 a - 8 \cos^2 a + 1.$$

$$\cos 5\,a = 16 \cos^5 a - 20 \cos^3 a + 5 \cos a.$$

$$\cos 6\,a = 32 \cos^6 a - 48 \cos^4 a + 18 \cos^2 a - 1.$$

$$\tan 2\,a = \frac{2 \tan a}{1 - \tan^2 a}.$$

$$\operatorname{ctn} 2\,a = \frac{\operatorname{ctn}^2 a - 1}{2 \operatorname{ctn} a}.$$

$$\sin (a \pm \beta) = \sin a \cdot \cos \beta \pm \cos a \cdot \sin \beta.$$

$$\cos (a \pm \beta) = \cos a \cdot \cos \beta \mp \sin a \cdot \sin \beta.$$

$$\tan (a \pm \beta) = \frac{\tan a \pm \tan \beta}{1 \mp \tan a \cdot \tan \beta}.$$

$$\operatorname{ctn} (a \pm \beta) = \frac{\operatorname{ctn} a \cdot \operatorname{ctn} \beta \mp 1}{\operatorname{ctn} a \pm \operatorname{ctn} \beta}.$$

$$\sin a \pm \sin \beta = 2 \sin \tfrac{1}{2} (a \pm \beta) \cdot \cos \tfrac{1}{2} (a \mp \beta).$$

$$\cos a + \cos \beta = 2 \cos \tfrac{1}{2} (a + \beta) \cdot \cos \tfrac{1}{2} (a - \beta).$$

$$\cos a - \cos \beta = - 2 \sin \tfrac{1}{2} (a + \beta) \cdot \sin \tfrac{1}{2} (a - \beta).$$

$$\tan a \pm \tan \beta = \frac{\sin (a \pm \beta)}{\cos a \cdot \cos \beta}.$$

$$\operatorname{ctn} a \pm \operatorname{ctn} \beta = \pm \frac{\sin (a \pm \beta)}{\sin a \cdot \sin \beta}.$$

$$\sin \alpha \cos \beta = \tfrac{1}{2} \left[\sin (\alpha + \beta) + \sin (\alpha - \beta) \right]$$

$$\cos \alpha \cos \beta = \tfrac{1}{2} \left[\cos (\alpha + \beta) + \cos (\alpha - \beta) \right]$$

$$\sin \alpha \sin \beta = \tfrac{1}{2} \left[\cos (\alpha - \beta) - \cos (\alpha + \beta) \right]$$

$$\frac{d(uv)}{dx} = v\frac{du}{dx} + u\frac{dv}{dx}.$$

$$\frac{d\left(\dfrac{u}{v}\right)}{dx} = \frac{v\dfrac{du}{dx} - u\dfrac{dv}{dx}}{v^2}.$$

$$\frac{d f(u)}{dx} = \frac{d f(u)}{du} \cdot \frac{du}{dx}.$$

$$\frac{d^2 f(u)}{dx^2} = \frac{df}{du} \cdot \frac{d^2 u}{dx^2} + \frac{d^2 f}{du^2} \cdot \frac{du^2}{dx^2}.$$

$$\frac{dx^n}{dx} = nx^{n-1}.$$

$$\frac{de^x}{dx} = e^x.$$

$$\frac{da^u}{dx} = a^u \cdot \frac{du}{dx} \cdot \log_e a.$$

$$\frac{d \sin x}{dx} = \cos x.$$

$$\frac{d \cos x}{dx} = -\sin x.$$

$$\frac{d \tan x}{dx} = \sec^2 x.$$

$$\frac{d \operatorname{ctn} x}{dx} = -\csc^2 x.$$

$$\frac{d \sec x}{dx} = \tan x \cdot \sec x.$$

$$\frac{d \csc x}{dx} = -\operatorname{ctn} x \cdot \csc x.$$

$$\frac{d \sin^{-1} x}{dx} = \frac{1}{\sqrt{1-x^2}}.$$

$$\frac{d \cos^{-1} x}{dx} = \frac{-1}{\sqrt{1-x^2}}.$$

$$\frac{d \tan^{-1} x}{dx} = \frac{1}{1+x^2}.$$

$$\frac{d \operatorname{ctn}^{-1} x}{dx} = -\frac{1}{1+x^2}.$$

$$\frac{d \sec^{-1} x}{dx} = \frac{1}{x\sqrt{x^2-1}}.$$

$$\frac{d \csc^{-1} x}{dx} = -\frac{1}{x\sqrt{x^2-1}}.$$

$$\frac{d \sinh x}{dx} = \cosh x.$$

$$\frac{d \cosh x}{dx} = \sinh x.$$

$$\frac{d \tanh x}{dx} = \operatorname{sech}^2 x.$$

$$\frac{d \operatorname{ctnh} x}{dx} = -\operatorname{csch}^2 x.$$

$$\frac{d \operatorname{sech} x}{dx} = -\operatorname{sech} x \cdot \tanh x.$$

$$\frac{d \operatorname{csch} x}{dx} = -\operatorname{csch} x \cdot \operatorname{ctnh} x$$

$$\frac{d \sinh^{-1} x}{dx} = \frac{1}{\sqrt{x^2+1}}.$$

$$\frac{d \cosh^{-1} x}{dx} = \frac{1}{\sqrt{x^2-1}}.$$

$$\frac{d \tanh^{-1} x}{dx} = \frac{1}{1-x^2}.$$

$$\frac{d \operatorname{ctnh}^{-1} x}{dx} = \frac{1}{1-x^2}.$$

$$\frac{d \operatorname{sech}^{-1} x}{dx} = \frac{-1}{x\sqrt{1-x^2}}.$$

$$\frac{d \operatorname{csch}^{-1} x}{dx} = \frac{-1}{x\sqrt{x^2+1}}.$$

$$\frac{d}{db}\int_a^b f(x)\, dx = f(b).$$

$$\frac{d}{da}\int_a^b f(x)\, dx = -f(a).$$

Index

NOTES

NOTES

NOTES

www.ingramcontent.com/pod-product-compliance
Lightning Source LLC
Chambersburg PA
CBHW081241180526
45171CB00005B/503